Motion Simulation and Mechanism Design Using SOLIDWORKS Motion 2024

Kuang-Hua Chang, Ph.D.

School of Aerospace and Mechanical Engineering
The University of Oklahoma
Norman, OK

PUBLICATIONS

SDC Publications
P.O. Box 1334
Mission, KS 66222
913-262-2664
www.SDCpublications.com

Examination Copies

Books received as examination copies are for review purposes only and may not be made available for student use. Resale of examination copies is prohibited.

Electronic Files

Any electronic files associated with this book are licensed to the original user only. These files may not be transferred to any other party.

Trademarks

© Dassault Systèmes. All rights reserved. **3D**EXPERIENCE, the 3DS logo, the Compass icon, IFWE, 3DEXCITE, 3DVIA, BIOVIA, CATIA, CENTRIC PLM, DELMIA, ENOVIA, GEOVIA, MEDIDATA, NETVIBES, OUTSCALE, SIMULIA and SOLIDWORKS are commercial trademarks or registered trademarks of Dassault Systèmes, a European company (Societas Europaea) incorporated under French law, and registered with the Versailles trade and companies registry under number 322 306 440, or its subsidiaries in the United States and/or other countries.

All statements are strictly based on the author's opinion. Dassault Systèmes and its affiliates disclaim any liability, loss, or risk incurred as a result of the use of any information or advice contained in this book, either directly or indirectly.

SOLIDWORKS®, eDrawings®, SOLIDWORKS Simulation®, SOLIDWORKS Flow Simulation, and SOLIDWORKS Sustainability are a registered trademark of Dassault Systèmes SOLIDWORKS Corporation in the United States and other countries; certain images of the models in this publication courtesy of Dassault Systèmes SOLIDWORKS Corporation.

The publisher and the author make no representations or warranties with respect to the accuracy or completeness of the contents of this work and specifically disclaim all warranties, including without limitation warranties of fitness for a particular purpose. No warranty may be created or extended by sales or promotional materials. Dimensions of parts are modified for illustration purposes. Every effort is made to provide an accurate text. The authors and the manufacturers shall not be held liable for any parts, components, assemblies or drawings developed or designed with this book or any responsibility for inaccuracies that appear in the book. Web and company information was valid at the time of this printing.

The Y14 ASME Engineering Drawing and Related Documentation Publications utilized in this text are as follows: ASME Y14.1 1995, ASME Y14.2M-1992 (R1998), ASME Y14.3M-1994 (R1999), ASME Y14.41-2003, ASME Y14.5-1982, ASME Y14.5-1999, and ASME B4.2. Note: By permission of The American Society of Mechanical Engineers, Codes and Standards, New York, NY, USA. All rights reserved.

Download all needed model files from the SDC Publication website (www.SDCpublications.com/downloads/978-1-63057-645-5).

ISBN-13: 978-1-63057-645-5

ISBN-10: 1-63057-645-X

Printed and bound in the United States of America.

Preface

This book is written to help you become familiar with *SOLIDWORKS Motion* (also called *Motion* in this book), which is an add-in module of the *SOLIDWORKS* software. *Motion* supports users to conduct modeling and analysis (or simulation) of mechanisms in a virtual (computer) environment. Capabilities in *Motion* allow you to use solid models created in *SOLIDWORKS* to simulate and visualize mechanism motion and system performance. Using *Motion* early in the product development stage could prevent costly (and often painful) redesign due to design defects found in the physical testing phase. Therefore, in general, using *Motion* for support of design decision making contributes to a more cost effective, reliable, and efficient product design process.

This book covers basic concepts and frequently used commands required to advance readers from a novice to an intermediate level in using *SOLIDWORKS Motion*. Basic concepts discussed in this book include model generation, such as creating assembly mates for a proper motion analysis model; carrying out simulation and animation; and visualizing simulation results, such as graphs and spreadsheet data. These concepts are introduced using simple yet realistic examples.

Verifying results obtained from computer simulation is extremely important. One of the unique features of this book is the incorporation of theoretical discussions for kinematic and dynamic analyses in conjunction with the simulation results obtained using *Motion*. The purpose of the theoretical discussions lies in solely supporting the verification of simulation results, rather than providing an in-depth discussion on the subject of mechanism design. *SOLIDWORKS Motion* is not foolproof. It requires a certain level of experience and expertise to master the software. Before arriving at that level, it is critical for you to verify simulation results whenever possible. Verifying simulation results will increase your confidence in using the software and prevent you from being fooled (hopefully, only occasionally) by erroneous simulation results produced by the software.

Example files have been prepared for you to go through all lessons, including *SOLIDWORKS* parts and assemblies. Completed *Motion* simulation models are also provided for your reference. You may want to start each lesson by reviewing the first sections (that is; overview of the lesson, and the motion example) and opening the assembly model with complete results in *SOLIDWORKS Motion* to see the motion simulation, in hopes of gaining more understanding about the example problems. In addition, *Excel* spreadsheets that support the theoretical calculations of selected examples are also available. You may download all model files and *Excel* files from the *SDC Publications* web site:

https://www.sdcpublications.com/downloads/978-1-63057-645-5

This book is written following a project-based learning approach and is intentionally kept simple to help you learn *Motion* more efficiently. Therefore, it is not intended in this book to include every single detail about *Motion*. For a complete reference of *Motion*, you may use on-line help in *SOLIDWORKS*, or visit the *SOLIDWORKS Corporation* web site at:

http://www.solidworks.com/

This book should serve self-learners well. If such describes you, you are expected to have a basic Physics and Mathematics background, preferably a bachelor's or associate degree in science or engineering. In addition, this book assumes that you are familiar with the basic concept and operation of *SOLIDWORKS* part and assembly modes. A self-learner should be able to complete all lessons in this book in about fifty hours. An investment of fifty hours should advance you from a novice to an intermediate level user. Not a bad investment at all!!

This book also supports class instruction. The book will most likely be used as a supplemental textbook for courses like *Mechanism Design*, *Rigid Body Dynamics*, *Computer-Aided Design*, *Computer-Aided Engineering*, *Design Principles*, or *Capstone Design*. This book should cover four to six weeks of class instruction, depending on how the courses are taught and the technical background of the students. Some of the exercise problems provided at the end of the lessons may require significant effort for students to complete. The author strongly encourages instructors and/or teaching assistants to go through those exercises before assigning them to students.

Examples presented in this book have been created and verified using *SOLIDWORKS 2024 SP1.0*. If you are using a different *SOLIDWORKS* version (or service pack), you may see a few minor differences in the menu selections and options, which should not hinder you from learning the software.

One last note: numerous software bugs were discovered in 2024 SP1.0 (among them, some were found in earlier versions and reported to the *SOLIDWORKS* software team). Some bugs lead to incorrect simulation results. Some bugs, in this book, were avoided by creating "get-around" models. All have been clearly explained in individual lessons, where they were found. Readers are encouraged to take extra caution while going over these lessons.

KHC
Norman, Oklahoma
March 4, 2024

About the Author

Dr. Kuang-Hua Chang is a professor for the School of Aerospace and Mechanical Engineering at the University of Oklahoma (OU), Norman, OK. He received his diploma in Mechanical Engineering from the National Taipei Institute of Technology, Taiwan, in 1980; and M.S. and Ph.D. degrees in Mechanical Engineering from the University of Iowa in 1987 and 1990, respectively. Since then, he joined the Center for Computer-Aided Design (CCAD) at Iowa as a Research Scientist and shortly after CAE Technical Area Manager. In 1997, he joined OU. He teaches mechanical design and manufacturing, in addition to conducting research in computer-aided modeling and simulation for design and manufacturing of mechanical systems. His work has been published in 10 books and more than 150 articles in international journals and conference proceedings. He has also served as a technical consultant to US industry and foreign companies, including LG-Electronics, Seagate Technology, etc.

Table of Contents

Lesson 6: A Spring-Mass System

Lesson 7: A Slider-Crank Mechanism

Lesson 8: A Slider-Crank Mechanism—Event-Based Dynamic Simulation

Lesson 9: A Rail Carriage Example

Lesson 10: A Compound Spur Gear Train

Lesson 11: Planetary Gear Train Systems

Lesson 12: Cam and Follower

Lesson 13: Simulation of Recreational Waterslide

Lesson 14: Kinematic Analysis of a Racecar Suspension

Appendix A: Defining Joints

Lesson 1: Introduction to *SOLIDWORKS Motion*

1.1 Overview of the Lesson

This lesson intends to provide you with a brief overview on *SOLIDWORKS Motion*. *SOLIDWORKS Motion*, formerly called *COSMOSMotion* (*SOLIDWORKS* 2008 and before), is a virtual prototyping tool that supports animation, analysis and design of mechanisms. Instead of building and testing physical prototypes of the mechanism, you may use *SOLIDWORKS Motion* (also called *Motion* in this book) to evaluate and refine the mechanism before finalizing the design and entering the functional prototyping stage. *Motion* will help you analyze and eventually design better engineering products that involve moving parts. More specifically, the software enables you to size motors and actuators, determine power consumption, layout linkages, develop cams, understand gear drives, size springs and dampers, and determine how contacting parts behave, which would usually require tests of physical prototypes. With such information, you will gain insight into how the mechanism works and why it behaves in certain ways. You will be able to modify the design and often achieve better design alternatives by using the more convenient and less expensive virtual prototypes. In the long run, using virtual prototyping tools, such as *Motion*, will help you become a more experienced and competent design engineer by gaining insights into the mechanics of parts interaction in mechanisms.

In this lesson, we will start with a brief introduction to *Motion* and discuss various types of physical problems that *Motion* is capable of solving. We will then discuss capabilities offered by *Motion* for creating motion models, conducting motion analyses, and viewing motion analysis results. In the final section of this lesson, we will preview examples employed in this book and topics to learn by going over these examples.

Note that materials presented in this lesson will be kept brief. More details on various aspects of the mechanism design and analysis using *Motion* will be provided in later lessons.

1.2 What is *SOLIDWORKS Motion*?

SOLIDWORKS Motion is a computer software tool that supports engineers to analyze and design mechanisms. It is a module of the *SOLIDWORKS* product family developed by *Dassault Systèmes SOLIDWORKS Corporation*. This software supports users to create virtual mechanisms that answer general questions in product design as described next. A single piston engine shown in Figures 1-1 and 1-2 is used to illustrate some typical questions of mechanism designs, such as the following:

1. Will the components of the mechanism collide or interfere in operation? For example, will the connecting rod collide with the inner surface of the piston or the inner surface of the engine case during operation?

2.   Will the components in the mechanism you are designing move according to your intent? For example, will the piston stay entirely in the piston sleeve? Will the system lock up when the firing force aligns vertically with the connecting rod?

3.   How much torque or force does it take to drive the mechanism? For example, what will be the minimum firing load to move the piston? Note that in this case, proper friction forces must be added to simulate the resistance between parts in contact in the mechanism before a realistic firing force can be calculated.

4.   How fast will the components move, e.g., the linear velocity of the piston?

5.   What is the reaction force or torque generated at a connection (also called *joint* or *constraint*) between components (or bodies) during motion? For example, what is the reaction force at the joint between the connecting rod and the piston pin? This reaction force is critical since the structural integrity of the piston pin and the connecting rod must be ensured, i.e., they must be structurally strong and durable enough to sustain the firing load in operation.

The modeling and analysis capabilities in *Motion* will help you answer these common questions accurately and realistically, as long as the motion model is properly defined.

Figure 1-1 A Single Piston Engine
(Unexplode View)

Figure 1-2 Single Piston Engine (Explode View)

The capabilities available in *Motion* also help you search for better design alternatives. A better design alternative is very much problem dependent. It is critical that a design objective be clearly defined by the designer up front before searching for better design alternatives. For the engine example, a better design alternative could be a design that reveals:

1.   A smaller reaction force applied to the connecting rod, and
2.   No collisions or interference between components.

One of the common approaches for searching for design alternatives is to vary the component sizes of the mechanism. In order to vary component sizes for exploring better design alternatives, the parts and assembly must be adequately parameterized to capture design intents. At the parts level, design parameterization implies creating solid features and relating dimensions so that when a dimension value is changed, the part can be rebuilt properly. At the assembly level, design parameterization involves defining assembly mates and relating dimensions across parts. When an assembly is fully parameterized, a change in dimension value can be propagated automatically to all parts affected. Parts affected must be rebuilt successfully, and at the same time they will have to maintain proper position and orientation with respect to one another without violating any assembly mates or revealing part penetration or excessive gaps. For example, in this single-piston engine assembly, a change in the bore diameter of the engine case will alter not only the geometry of the engine case, but also all other parts affected, such as the piston, piston sleeve, and even the crankshaft, as illustrated in Figure 1-3. Moreover, they all have to be rebuilt properly and the entire assembly must stay intact through assembly mates.

(a) Bore Diameter 1.2 inch   (b) Bore Diameter 1.6 inch

Figure 1-3 A Single-Piston Engine—Explode View

1.3   Mechanism Design and Motion Analysis

A mechanism is a mechanical device that transfers motion and/or force from a source to an output. It can be an abstraction (simplified representation) of a mechanical system represented as a linkage. A linkage consists of links (also called bodies), which are connected by joints (or connections), such as a revolute joint, to form open or closed chains (or loops, see Figure 1-4). Such kinematic chains, with at least one link fixed, become mechanisms. In this book, all links are assumed rigid. In general, a mechanism can be represented by its corresponding schematic sketch for discussion and presentation purposes. For example, a slider-crank mechanism characterizes the engine motion kinematically, as shown in Figure 1-5, which is a closed loop mechanism.

In general, there are two basic types of motion problems that you may have to solve in order to answer questions regarding mechanism analysis and design: kinematic and dynamic.

Kinematics is the study of motion without regard for the forces that cause the motion. A kinematic mechanism can be driven by a servomotor so that the position, velocity, and acceleration of each link of the mechanism can be analyzed at any given time. Typically, a kinematic analysis is conducted before the dynamic behavior of the mechanism can be simulated properly.

Dynamic analysis is the study of motion in response to externally applied loads. The dynamic behavior of a mechanism is governed by Newton's laws of motion. The simplest dynamic problem is the particle dynamics introduced in sophomore Dynamics class—for example, a spring-mass-damper system shown in Figure 1-6. In this case, motion of the mass is governed by the following equation derived from Newton's second law,

$$\sum F = p(t) - kx - c\dot{x} = m\ddot{x} \qquad (1.1)$$

where the over-dot ($\cdot$) represents the derivative with respect to time, double-dot ($\cdot\cdot$) represents second derivative, m is the total mass of the block, k is the spring constant, and c is the damping coefficient.

(a) Open Loop Mechanism   (b) Closed Loop Mechanism

Figure 1-4 General Mechanisms

Figure 1-5 Schematic View of the Engine Motion Model

For a rigid body, mass properties (such as the total mass, center of mass, mass moment of inertia, etc.) are taken into account for dynamic analysis. For example, motion of a pendulum shown in Figure 1-7 is governed by the following equation of motion,

$$\sum M = -mg \, sin\theta = I\ddot{\theta} \qquad (1.2)$$

where M is the external moment (or torque), I is the mass moment of inertia of the pendulum ($I = m\ell^2$), m is the pendulum mass, g is the gravitational acceleration, and $\ddot{\theta}$ is the angular acceleration of the pendulum.

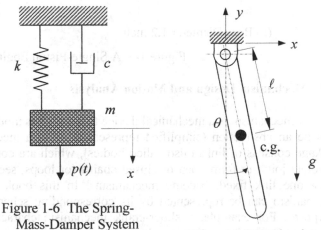

Figure 1-6 The Spring-Mass-Damper System

Figure 1-7 A Simple Pendulum

Dynamic analysis of a rigid body system, such as the single piston engine shown in Figure 1-3, is a lot more complicated than the single body problems. Usually, a system of differential and algebraic equations governs the motion and the dynamic behavior of the system. Newton's laws must be obeyed by individual bodies in the system at all times. The motion of the system will be determined by the loads acting on the bodies or joint axes (e.g., a torque driving the system). Reaction forces and moments at the joints hold the bodies together.

Note that in *Motion*, you may create a kinematic analysis model; e.g., using a motion driver (linear or rotary motor) to drive the mechanism; and then carry out dynamic analyses. In dynamic analysis, the inertia of the bodies will be taken into account for analysis; therefore, reaction forces will be calculated between bodies.

1.4 *SOLIDWORKS Motion* Capabilities

Overall Process

The overall process of using *SOLIDWORKS Motion* for analyzing a mechanism consists of three main steps: model generation, analysis (or simulation), and result visualization (or post-processing), as illustrated in Figure 1-8. Key entities that constitute a motion model include servomotors that drive the mechanism for kinematic analysis, external loads (force and torque), force entities such as spring and damper, and the initial conditions of the mechanism. Most importantly, assembly mates must be properly defined for the mechanism so that the motion model captures essential characteristics and accurately resembles the behavior of the physical mechanism.

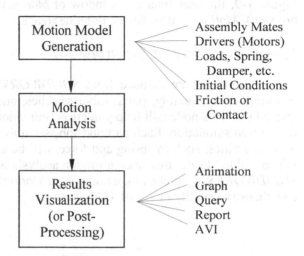

Figure 1-8 General Process of Using *SOLIDWORKS Motion*

The analysis or simulation capabilities in *Motion* employ simulation engine *ADAMS/Solver*, which solves the equations of motion for the mechanism. *ADAMS/Solver* calculates the position, velocity, acceleration, and reaction forces acting on individual moving parts in the mechanism. Typical simulation problems, including static (equilibrium configuration) and motion (kinematic and dynamic) are supported. More details about the analysis capabilities in *Motion* will be discussed later in this lesson.

The analysis results can be visualized in various forms. You may animate the motion of the mechanism or generate graphs for more specific information, such as the reaction force of a joint in time domain. You may also query results at specific locations for a given time. Furthermore, you may ask for a report on the results of your choices, such as the acceleration of a moving part in the time domain. You may also save the motion animation to an AVI for faster viewing and/or showing motion animation on other computers without using *SOLIDWORKS Motion*.

Operation Mode

Motion is embedded in *SOLIDWORKS*. It is indeed an add-in module of *SOLIDWORKS*. Transition from *SOLIDWORKS* to *Motion* is seamless. All the solid parts, materials, assembly mates, etc. defined in *SOLIDWORKS* are automatically carried over into *Motion*. *Motion* can be accessed through menus and windows inside *SOLIDWORKS*. The same assembly created in *SOLIDWORKS* is directly employed for creating motion models in *Motion*. In addition, part geometry is essential for mass property computations in motion analysis. In *Motion*, all mass properties calculated in *SOLIDWORKS* are ready for use. In addition, the detailed part geometry supports interference check (as well as detecting contact between bodies) for the mechanism during motion simulation in *Motion*.

User Interfaces—MotionManager Window

User interface of *Motion* is embedded in *SOLIDWORKS*. *SOLIDWORKS* users should find that it is straightforward to maneuver in *Motion*. The main interface for using *Motion* is through the *MotionManager* window, as shown in Figure 1-9. MotionManager is a separate window that is used to create motion models and conduct motion analyses. When you open an existing assembly (or part) in *SOLIDWORKS*, the *Motion Study* tab (with a default name *Motion Study 1*) will appear at the bottom of the graphics area. Clicking the *Motion Study* tab will bring up the *MotionManager* window. As shown in Figure 1-9, the user interface window of *MotionManager* consists of *MotionManager* tree (or *Motion* browser), *Motion* toolbar, filters, timeline area, etc.

MotionManager Tree (or Motion Browser)

Components are mapped from *SOLIDWORKS* assembly into *MotionManager* tree automatically, including root assembly, parts, sub-assemblies, and mates. An *Orientation and Camera Views* node is also added. This node will help you create animation in a specific view as desired, but has less to do with the motion simulation. Each part and sub-assembly can be expanded to show its composing components. Motion entities, such as spring and force, will be added to the tree once they are created. A result node will be added to the tree once a motion analysis is completed and result graphs are created. Similar to *SOLIDWORKS*, right clicking a node in the *MotionManager* tree will bring up command options that you can choose to modify or adjust *Motion* entities.

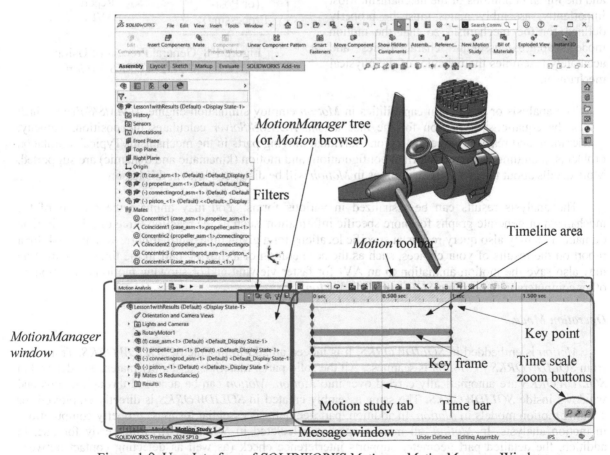

Figure 1-9 User Interface of *SOLIDWORKS Motion—MotionManager* Window

Motion Toolbar

The *Motion* toolbar shown in Figure 1-9 (and more detailed in Figure 1-10) provides major functions required to create and modify motion models, create and run analyses, and visualize results. The toolbar includes a type of motion study selection (for choosing from animation, basic motion, or motion analysis), calculate, animation controls, playback speed and mode options, save options, animation wizard, key point controls, simulation elements, results and graphs, and motion study property. As you move the mouse pointer over a button of the toolbar, a brief description about the functionality of the button appears. For example, if you move the mouse pointer close to the *Spring* button, a brief description about this button will appear next to it as well as in the *Message* window, located at the bottom left corner of the *MotionManager* window, as shown in Figure 1-9. The buttons in *Motion* toolbar and their functions are summarized in Table 1-1.

Figure 1-10 The *Motion* Toolbar

Note that you may choose *Animation*, *Basic Motion*, or *Motion Analysis* for a motion study. You can use *Animation* to animate simple operation of assemblies, such as rotation, zoom in/out, explode/collapse assembly, etc. You may also add motors to animate simple kinematic motion of the assembly.

Basic Motion supports you for approximating the effects of motors, springs, collisions, and gravity on assemblies. *Basic Motion* takes mass into account in calculating motion. The computation is relatively fast, so you can use this for creating presentation-worthy animations using physics-based simulations.

Motion Analysis (only available when *SOLIDWORKS Motion* is added in) provides you with accurate simulation and analysis on the effects of motion entities (including forces, springs, dampers, and friction) on an assembly. *Motion Analysis* uses computationally strong kinematic solvers, and accounts for material properties as well as mass and inertia in the computations. You can also use *Motion Analysis* to plot simulation results for further analysis.

As a general rule of thumb, you may use *Animation* to create presentation-worthy animations for motion that does not require accounting for mass or gravity; use *Basic Motion* to create presentation-worthy approximate simulations of motion that account for mass, collisions, or gravity; and use *Motion Analysis* to run computationally strong simulations that take the physics of the assembly motion into consideration. *Motion Analysis* is the most computationally intensive of the three options.

Both *Animation* and *Basic Motion* are available in core *SOLIDWORKS*. *Motion Analysis* is only available with the *SOLIDWORKS Motion* add-in to *SOLIDWORKS*. Before using this book, you are encouraged to check with your IT staff or system administrator to make sure the *Motion* module has been installed on your computer.

Filters

Filters can be used to limit the *MotionManager* tree by including only animated, driving, selected or simulation results. Filters help make a cleaner display of entities in the *MotionManager* tree.

Timeline Area

The area to the right of the *MotionManager* tree is the timeline area, which is the temporal interface for animation. The timeline area displays the times and types of animation events in the motion study.

Table 1-1 The Function Buttons in *Motion* Toolbar

Symbol	Name	Function
	Calculate	Calculates the current simulation. If you alter the simulation, you must recalculate it before replaying.
	Play from start	Reset components and play the simulation. Use after simulation has been calculated.
	Play	Play the simulation beginning at the current timebar location.
	Stop	Stop the animation.
	Playback mode: Normal	Play from beginning to end once.
	Playback mode: Loop	Continuously play, from beginning to end, then loop to beginning and continue playing.
	Playback mode: Reciprocal	Continuously play, from beginning to end, then reverse—play from end to beginning.
	Save animation	Save the animation as an AVI movie file.
	Animation wizard	Allow to quickly create rotate model, explode or collapse animation.
	Auto key	Click to automatically place a new key when you move or change components. Click again to toggle this option.
	Add/update key	Click to add a new key or update properties of an existing key.
	Motor	Create a motor for motion analysis.
	Spring	Add a spring between two components.
	Damper	Add a damper between two components.
	Force	Create a force for motion analysis.
	Contact	Create a 3D contact between selected components.
	Gravity	Add gravity to the motion study.
	Results and graphs	Create results and graphs.
	Motion study properties	Define motion study solution parameters.
	Simulation setup	Set up stress analysis on one or more components.
	Event-based motion view	Switch from the timeline view to an event-based view.
	Timeline view	Switch from event-based view to a timeline view.
	Collapse *MotionManager*	Collapse the *MotionManager* window.

The timeline area is divided by vertical grid lines (called time bar) corresponding to numerical markers showing the time. The numerical markers start at 00:00:00. You may click and drag a key to define the beginning and/or end time of the animation or motion simulation.

After a simulation is completed, you will see several horizontal bars appear in the timeline area. They are *Key Frames* in *Motion*. The *key frame* is the area between key points, which can be any length of time. It defines the time frame in which assembly component motion or visual property changes take place. A *Key Point* is the entity that corresponds to defined assembly component positions, visual properties, or simulation element states. You can use key points to represent a beginning or end of a change in animation position or other attributes at a given time. Whenever you position a new key point, it corresponds to motion or to changes to visual attributes. More about the timeline area will be discussed in *Lesson 2*.

Switching back and forth between *Motion* and *SOLIDWORKS* is straightforward. All you have to do is to click the *Model* tab (back to *SOLIDWORKS*) and *Motion Study* tab (to *Motion*) at the bottom of the graphics area.

Defining Motion Entities

The basic entities of a valid *Motion* simulation model consist of ground parts (or ground body), moving parts (or moving bodies), constraints (imposed by mates in *SOLIDWORKS* assembly), initial conditions (usually position and velocity of a moving body), and forces and/or drivers. Each of the basic entities will be briefly discussed next. More details can be found in later lessons.

Ground Parts (or Ground Body)

A ground part, or a ground body, represents a fixed reference in space. The first component brought into the assembly is usually stationary, therefore becoming a ground part. Parts (or sub-assemblies) assembled to the stationary components without any possibility to move become part of the ground body. A symbol *(f)* is placed in front of stationary components in the browser. You will have to identify moving and non-moving parts in your mechanism design, and use assembly mates to properly constrain the movement of the moving parts and completely fix the motion of non-moving parts.

Moving Parts (or Bodies)

A moving part (or a body) is an entity representing a single rigid component (or link) that moves relatively to other parts (or bodies). A moving part may consist of a single *SOLIDWORKS* part or a sub-assembly composed of multiple parts. When a sub-assembly is designated as a moving part, although the sub-assembly is movable, none of its composing parts is allowed to move relative to one another within the sub-assembly.

A moving body has six degrees of freedom: three translational and three rotational (shown in Figure 1-11), while a ground body has none. That is, a moving body can translate and rotate along the X-, Y-, and Z-axes of a coordinate system. Rotation of a rigid body is measured by referring the orientation of its local coordinate system to the global coordinate system, which is shown at the lower left corner on the graphics area in *SOLIDWORKS* (called *Reference Triad*). In *Motion*, the local coordinate system is assigned automatically, usually at the mass center of the part or sub-assembly. Mass properties, including total mass, mass moment of inertia, etc., are calculated using part geometry and material properties referring to the local coordinate system. A symbol *(-)* is placed in front of a moving component in the browser.

Tx: translational DOF along *X*-axis
Ty: translational DOF along *Y*-axis
Tz: translational DOF along *Z*-axis
Rx: rotational DOF along *X*-axis
Ry: rotational DOF along *Y*-axis
Rz: rotational DOF along *Z*-axis

Figure 1-11 Translational and Rotational Degrees of Freedom

Constraints

As mentioned earlier, an unconstrained rigid body in space has six degrees of freedom: three translational and three rotational. When you add a joint (or constraint) between two rigid bodies, you remove degrees of freedom between them. Since *SOLIDWORKS* 2008, more commonly employed joints, such as revolute, translation, cylindrical joint, etc., in *Motion* have been replaced by assembly mates, such as coincident and concentric. Like joints, assembly mates remove degrees of freedom between parts.

Each independent movement permitted by a constraint (either a joint or an assembly mate) is a free degree of freedom. The free degrees of freedom that a constraint allows can be translational or rotational along the three perpendicular axes, either global or local. For example, a concentric mate between the propeller assembly and the case of a single piston engine shown in Figure 1-12 allows one translational DOF (movement along the center axis, in this case, parallel to the *X*-axis of the global coordinate system, called reference triad in *SOLIDWORKS*) and one rotational DOF (rotating along *X*-axis). Since the case assembly is stationary, serving as the ground body, the propeller assembly has two free DOFs. Adding a coincident mate between the two respective mating faces of the engine case and the propeller shown in Figure 1-12 removes the remaining translational DOF, yielding a desired assembly that resembles the physical situation, i.e., with only the rotational DOF (along the *X*-axis).

In creating a motion model, instead of completely fixing all body motion, certain DOFs (translational and/or rotational) are left to allow for designated movement called free DOF. Such a movement will be either driven by a motor (for kinematic analysis) or determined by a force in dynamic analysis. For example, a rotary motor is created to drive the rotational DOF of the propeller in the engine example. This rotary motor rotates the propeller at a prescribed angular velocity. In addition to velocity, you may use the motor to drive a DOF at a prescribed displacement or acceleration, either translation (using a linear motor) or rotational (using a rotary motor). You may apply a force or a torque to a free translational or rotational DOF, respectively, for a dynamic simulation.

It is extremely important to understand assembly mates in order to create working motion models that correctly simulate the physical behavior of the mechanism. In addition to standard mates such as concentric and coincident (see Table 1-2 for a list of *SOLIDWORKS* standard mates), *SOLIDWORKS* provides advanced mates and mechanical mates (see Tables 1-3 and 1-4, respectively). Advanced mates offer additional ways to constrain or couple movements between bodies. For example, a linear coupler allows the motion of a translational or a rotational DOF of a given mate to be coupled to the motion of another mate. Also, the rotation motion of the propeller may be coupled to the translational motion of the piston (certainly if it makes sense). A coupler removes one additional degree of freedom from the motion model. Also, path mate (one of the advanced mates in *SOLIDWORKS*) allows a part to move along a curve slot, a groove, or fluting, varying its moving direction specified by the path curve. Such a capability supports animation and motion analysis of a different set of applications involving curvilinear motion. We will learn path mate in *Lesson 9*.

Mechanical mates include cam follower, gear, hinge, rack and pinion, screw, and universal joint. These are essential for motion models yet extremely easy to create in *Motion*. Some of the mechanical mates will be further discussed in later lessons (for examples Lessons 10 to 12).

In addition to mates, *Motion* provides solid body contact that helps to simulate physical problems involving contact between bodies. Essentially, solid body contact applies a force to separate the parts when they are in contact and prevent them from penetrating each other. The solid body contact will become active as soon as the parts are touching (or about to touch) each other. We will learn to create a contact constraint in *Lesson 3* and *Lesson 13*.

Figure 1-12 Assembly Mates Defined for the Propeller (Explode View)

Table 1-2 Standard Mates of *SOLIDWORKS* Assembly

Standard Mates	Descriptions
Coincident ⦦	Positions selected faces, edges, and planes (in combination with each other or combined with a single vertex) so they share the same infinite plane. Positions two vertices so they touch.
Parallel ⦥	Places the selected items so they remain a constant distance apart from each other.
Perpendicular ⊥	Places the selected items at a 90° angle to each other.
Tangent ♂	Places the selected items tangent to each other (at least one selection must be a cylindrical, conical, or spherical face).
Concentric ◎	Places the selections so that they share the same center line.
Lock 🔒	Maintains the position and orientation between two components.
Distance ⟷	Places the selected items with the specified distance between them.
Angle △	Places the selected items at the specified angle to each other.
Default	Place the first part to the default coordinate system in assembly.

Table 1-3 Advanced Mates of *SOLIDWORKS* Assembly

Advanced Mates	Descriptions
Symmetric	Forces two similar entities to be symmetric about a plane or planar face.
Width	Centers a tab within the width of a groove.
Path	Constrains a selected point on a component to a path.
Linear/Linear Coupler	Establishes a relationship between the translation of one component and the translation of another component.
Limit	Allows components to move within a range of values for distance and angle mates.
Profile Center	Automatically center-aligns geometric profiles to each other and fully defines the components.

Table 1-4 Mechanical Mates of *SOLIDWORKS* Assembly

Mechanical Mates	Descriptions
Cam	Forces a cylinder, plane, or point to be coincident or tangent to a series of tangent extruded faces.
Gear	Forces two components to rotate relative to one another about selected axes (or respective pitch circles).
Hinge	Limits the movement between two components to one rotational degree of freedom.
Rack and Pinion	Linear translation of one part (the rack) causes circular rotation in another part (the pinion), and vice versa.
Screw	Constrains two components to be concentric, and also adds a pitch relationship between the rotation of one component and the translation of the other.
Slot	Mates bolts to straight or arced slots and mate slots to slots. You can select an axis, cylindrical face, or a slot to create slot mates.
Universal Joint	The rotation of one component (the output shaft) about its axis is driven by the rotation of another component (the input shaft) about its axis.

Degree of Freedom Count

As mentioned earlier, an unconstrained body in space has six degrees of freedom. When assembly mates are added to assemble parts, constraints are imposed to eliminate DOFs accordingly so as to restrict the relative motion between them as desired.

Let us go back to the engine example shown in Figure 1-12. A concentric mate between the propeller and the engine case restricts movement on four DOFs (Ty, Tz, Ry, and Rz) so that only two movements are allowed, one translational (Tx) and one rotational (Rx). In order to restrict the translational movement, a coincident mate was added. A coincident mate between two respective mating faces of the propeller and the engine case restricts movement on three DOFs, i.e., Tx, Ry, and Rz. Even though combining these two mates gives desired rotational motion between the propeller and the case, there are redundant DOFs being restricted, i.e., Ry and Rz in this case.

It is important that you understand how to count the overall degrees of freedom for a motion model. For a given motion model, the number of degrees of freedom can be determined by using the Gruebler's count, defined as:

$$D = 6M - N - O \tag{1.3}$$

where D is the Gruebler's count representing number of free degrees of freedom of the mechanism, M is the number of moving bodies (excluding the ground body), N is the number of DOFs restricted by all mates, and O is the number of motion drivers (motors) defined in the system. For the motion model consisting of the propeller, the engine case, and the rotary motor, the Gruebler's count is:

$$D = 6 \times 1 - (4+3) - 1 = -2$$

However, we know that the propeller can only rotate along the X-axis; therefore, there is only one free DOF for the system (Rx). The count should be *1*. After adding the rotary motor, the count should become zero. The calculation gives us *-2* due to the fact that there are two redundant DOFs, Ry and Rz, which were restrained by both concentric and coincident mates. If we remove the redundant DOFs, the count becomes:

$$D = 6 \times 1 - (4+3-2) - 1 = 0$$

Another example, a door assembled to door frame using two hinge joints. Each hinge joint allows only one rotational movement along the axis of the hinge. The second hinge adds five redundant DOFs. The Gruebler's count becomes:

$$D = 6 \times 1 - 2 \times 5 = -4$$

Again, if we remove the redundant DOFs, the count becomes, as it should be:

$$D = 6 \times 1 - (2 \times 5 - 5) = 1$$

This is before any motor is added.

For kinematic analysis, the Gruebler's count must be equal to or less than *0* after adding motors. The *ADAMS/Solver* recognizes and deactivates redundant constraints during analysis. For a kinematic analysis, if you create a model and try to animate it with a Gruebler's count greater than *0*, the animation will not run and an error message will appear. For the door example, the vertical movement constrained by the second hinge is identified as redundant and removed from the solution. As a result, if conducting a dynamic simulation, the entire vertical force is carried by the first hinge. No (reaction) force will be calculated at the second hinge.

To get the Gruebler's count to zero, it is often required to replace mates that remove a larger number of constraints with mates that remove a smaller number of constraints and still restrict the mechanism motion in the same way. This is usually difficult if not entirely impossible. *Motion* detects the redundancies and ignores redundant DOFs in all analyses, except for dynamic analysis. In dynamic analysis, the redundancies lead to an outcome with a possibility of incorrect reaction results, yet the motion is correct. For complete and accurate reaction forces, it is critical that you eliminate redundancies from your mechanism if possible. The challenge is to find the mates that will impose non-redundant constraints and still allow for the intended motion. Very often this is not possible in *Motion* since instead of using kinematic joints (like revolute joint) it employs more primitive assembly mates for creating motion models. A concentric and a coincident mate in *SOLIDWORKS* is kinematically equivalent to a revolute joint, as illustrated in Figure 1-12 between the propeller and the engine case. A revolute joint removes five DOFs (with no redundancy); however, combining a concentric and a coincident mate removes seven DOFs, among which two are redundant. Using assembly mates to create motion models usually creates redundant DOFs.

The best strategy using *Motion* is to create an assembly that accurately resembles the physical mechanism by using assembly mates that capture the characteristics of the motion revealed in the physical application. That is, first you should focus on creating an assembly that correctly captures the kinematic behavior of the mechanism. If you are conducting a dynamic analysis and want to capture reaction forces at critical components, you will have to examine the assembly mates and identify redundant DOFs. Then, check reaction forces at all mates for the relevant component and only take the reaction forces that make sense; mostly non-zero forces since zero reaction force is usually reported at redundant DOFs by *Motion*.

Forces

Forces are used to operate a mechanism. Physically, forces are produced by motors, springs, dampers, gravity, etc. A force entity in *Motion* can be a force or torque. *Motion* provides three types of forces: applied forces, flexible connectors, and gravity.

Applied forces are forces that cause the mechanism to move in designated ways. Applied forces are very general, which involves defining the force direction and the force magnitude by specifying a constant force value or expression function, such as a harmonic function. The applied forces in *Motion* include action-only force or moment (where force or moment is applied at a point or an axis on a single rigid body, and no reaction forces are calculated), action and reaction force and moment, and impact force.

Flexible connectors resist motion and are simpler and easier to use than applied forces because you only supply constant coefficients for the forces, for instance a spring constant. The flexible connectors include translational springs, torsional springs, translational dampers, torsional dampers, and bushings.

A magnitude and a direction must be included for a force definition. You may select a predefined function, such as a harmonic function, to define the magnitude of the force or moment. For spring and damper, *Motion* automatically makes the force magnitude proportional to the distance or velocity, respectively, between two points, based on the spring constant and damping coefficient entered. The direction of a force (or moment) can be defined by either along an axis defined by an edge or along the line between two points, where a spring or a damper is defined.

Initial Conditions

In motion simulations, initial conditions consist of initial configuration of the mechanism and initial velocity of one or more components of the mechanism. Motion simulation must start with a properly assembled solid model that determines an initial configuration of the mechanism, composed by position and orientation of individual components. The initial configuration can be completely defined by assembly mates. However, one or more assembly mates will have to be suppressed, if the assembly is fully constrained, to provide adequate movement. In *SOLIDWORKS Motion*, initial velocity is assigned to a moving body (part or sub-assembly). The initial velocity can be translational or rotational along a direction spanned by the three axes.

Motion Drivers

Motion drivers (or motors) are used to impose a particular movement on a free DOF over time. A motion driver specifies position, velocity, or acceleration as a function of time, and can control either translational or rotational motion. When properly defined, motion drivers will account for the remaining DOFs of the mechanism that bring the Gruebler's count to zero (exactly zero after removing all redundant DOFs) for a kinematic analysis.

Motion Simulation

The *ADAMS/Solver* employed by *Motion* is capable of solving general problems, such as static (equilibrium configuration), kinematic, and dynamic. Three numerical solvers are provided. They are GSTIFF, SI2_GSTIFF, and WSTIFF. GSTIFF is the default (numerical) integrator and is fast and accurate for displacements. It is used for a wide range of motion simulations. SI2_GSTIFF provides better accuracy of velocities and accelerations but can be significantly slower. WSTIFF provides better accuracy for special problems, such as discontinuous forces.

Static analysis is used to find the rest position (equilibrium condition) of a mechanism, in which none of the bodies are moving. A simple example of the static analysis is illustrated in Figure 1-13, in which an equilibrium position of the block is to be determined according to its own mass *m*, the two spring constants k_1 and k_2, and the gravity *g*. Note that in *Motion*, you will have to include damping or friction for the mechanism to reach an equilibrium or rest configuration.

Figure 1-13 Static Analysis

Figure 1-14 Kinematic Analysis

Figure 1-15 Dynamic Analysis

As discussed earlier, kinematics is the study of motion without regard for forces or torque. A mechanism can be driven by a motion driver for a kinematic analysis, where the position, velocity, and acceleration of each link of the mechanism can be analyzed at any given time. Figure 1-14 shows such an example, where a servomotor drives a slider-crank mechanism at a constant angular velocity.

Dynamic analysis is employed for studying the mechanism motion in response to loads, as illustrated in Figure 1-15, in which a force *p(t)* is applied at the slider in the horizontal direction to cause the slider to move, leading to the movement of the connecting rod and the crank. Dynamic analysis is the most complicated and common, and sometimes a more time-consuming analysis.

Viewing Results

In *Motion*, results of the motion analysis can be visualized using animations, graphs, reports, and queries. Animations show the configuration of the mechanism in consecutive time frames. Animations will give you a global view on how the mechanism behaves; for example, the single-piston engine shown in Figure 1-16. You may also export the animation to AVI and insert the video clip into your PowerPoint slide for a technical presentation.

In addition, you may choose a joint (assembly mates in *Motion*) or a part to generate result graphs; for example, position vs. time of the mass center of the piston in the engine example shown in Figure 1-17. These graphs give you a quantitative understanding of the characteristics of the mechanism.

You may also query the results by moving the mouse pointer closer to the curve in a graph and leave the pointer for a short period. The result data will appear next to the pointer. In addition, you may ask *Motion* for a report that includes a complete set of results output in the form of textual data or a Microsoft® Excel spreadsheet.

In addition to the capabilities discussed above, *Motion* allows you to check interference between parts during motion (please see *Lesson 7* for more details). Furthermore, the reaction forces calculated can be used to support structural analysis using, for example, *SOLIDWORKS Simulation*.

Figure 1-16 Motion Animation

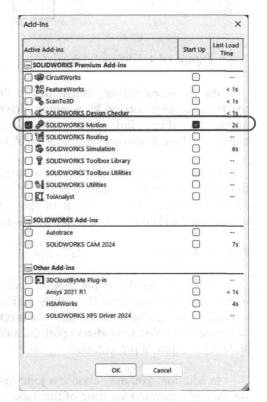

Figure 1-17 Graph of Position of the Piston vs. Time

1.5 Open Lesson 1 Model

A motion model for the single piston engine example shown in Figure 1-1 has been created for you. Download the files from www.sdcpublications.com, unzip them, and locate the engine model (filename: *Lesson1withResults.SLDASM*) under *Lesson 1*. Copy or move *Lesson 1* folder to your hard drive.

Start *SOLIDWORKS* and open assembly *Lesson1withResults.SLDASM*. You should see an assembled engine model similar to that of Figure 1-1. Also, the *Motion Study* tab (with a default name *Motion Study 1*) will appear at the bottom of the graphics area.

To use full capabilities of *Motion*, you will need to activate the *Motion* module by choosing from the pull-down menu

Tools → Add-Ins

In the *Add-Ins* window shown in Figure 1-18, click *SOLIDWORKS Motion* in both boxes (*Active Add-ins* and *Start Up*), and then click *OK*.

Figure 1-18 The *Add-Ins* Window

Click the *Motion Study* tab to bring up the *MotionManager* window. Note that a rotary motor, *RotaryMotor1*, has been added to the motion model.

Click the *Play* button on the *Motion* toolbar; you should see that the propeller starts rotating, similar to that of Figure 1-16. Also, you may want to hide *case_asm* to see the motion inside the case, especially between the connection rod and the piston. You may do so by right clicking *case_asm<1>* from the *MotionManager* tree and choosing *Hide*.

1.6 Motion Examples

More motion examples will be introduced in this book to illustrate the step-by-step details of modeling, simulation, and result visualization capabilities in *Motion*.

We will start with the same engine example in *Lesson 2*. *Lesson 2* introduces numerous animations and basic motion study capabilities that help you become familiar with the basic operations in *MotionManager* window, especially the timeline area. In *Lesson 3*, a simple ball-throwing example will be presented. This example provides you with a quick run-through on using *Motion*. This lesson will also involve creating solid body contact. *Lessons 4* through *12* focus on modeling and analysis of basic mechanisms and dynamic systems. In these lessons, you will learn assembly mates; forces and connections, including springs, gears, cam-followers; drivers and forces; various analyses; and graphs and results. Two application lessons are offered. *Lesson 13* presents a motion study of riding object (a small hemispheric object) on a recreational waterslide. *Lesson 14* illustrates the kinematic analysis of suspension of a SAE Formula style racecar. All examples and topics to be discussed in each lesson are summarized in the following table.

Table 1-5 Examples Employed in This Book

Lesson	Title	Motion Model	Problem Type	Things to Learn
2	Single Piston Engine Example		Animation and basic motion	1. This lesson focuses on helping you become familiar with the basic operations in *MotionManager* window. 2. You will learn animation and basic motion study capabilities, especially using *Animation Wizard* to create simple animations in assembly.
3	Ball Throwing Example		Particle Dynamics	1. This lesson offers a quick run-through of general modeling and simulation capabilities in *Motion*. 2. You will learn the general process of using *Motion* to construct a motion model, run simulation, and visualize motion simulation results. 3. Gravity will be turned on and a solid body contact will be defined between the ball and the ground to simulate a ball bouncing scenario. 4. Simulation results are verified using analytical equations of motion.

Table 1-5 Examples Employed in This Book (Cont'd)

Lesson	Title	Motion Model	Problem Type	Things to Learn
4	Simple Pendulum		Particle Dynamics	1. This lesson provides a discussion on using concentric and coincident mates to create a revolute joint that allows the simulation of pendulum motion using *Motion*. 2. Simulation results are verified using analytical equations of motion.
5	Friction of a Sliding Object		Particle Dynamics	1. This lesson provides a discussion on friction of a sliding object. 2. Both contact friction and joint friction offered in *Motion* are presented. 3. Simulation results are verified using analytical equations of motion.
6	Spring-Mass System		Particle Dynamics	1. This is a classical spring-mass system example you learned in Sophomore Dynamics class. 2. You will learn how to create a mechanical spring, align the block with the slope surface, and add an external force to pull the block. 3. Simulation results are verified using analytical equations learned in Dynamics.
7	Slider Crank Mechanism		Multibody Dynamic Analyses	1. This lesson uses a slider-crank mechanism to conduct kinematic and dynamic simulations. 2. You will learn to create motion drivers as well as add a firing force for simulation. 3. The interference checking capability will be discussed. 4. Kinematic analysis results are verified using analytical equations of motion.
8	Event-Based Dynamic Simulation		Multibody Dynamic Analyses	1. This lesson uses the same slider-crank mechanism to present event-based dynamic simulation. 2. You will learn to create task trigger and task action, from which a more realistic firing load is created for the simulation of the slider-crank mechanism.

Table 1-5 Examples Employed in This Book (Cont'd)

Lesson	Title	Motion Model	Problem Type	Things to Learn
9	Rail-Carriage Example		Path Mate	1. This lesson introduces a useful and powerful mate provided in *SOLIDWORKS*, the path mate. Such a capability supports animation and motion study of a different set of applications involving curvilinear motion.
10	Compound Spur Gear Train		Gear Train Analysis	1. This lesson focuses on simulating motion of a spur gear train system. 2. You will learn how to use *SOLIDWORKS* to create gear mates, and use *Motion* to simulate the gear train motion. 3. Gear rotation speed is verified using analytical calculation.
11	Planetary Gear Train Systems		Gear Train Analysis	1. This lesson focuses on simulating motion of planetary gear train systems. 2. You will learn how to use *SOLIDWORKS* to create gear mates and use *Motion* to simulate the gear train motion. 3. Both single planetary and multi-planetary gear train systems are included. 4. Gear rotation speed is verified using analytical calculations.
12	Cam and Follower		Multibody Dynamic Analysis	1. This lesson discusses cam and follower connection. 2. An inlet or outlet valve system of an internal combustion engine will be created and simulated. 3. Position and velocity of the valve will be graphed to monitor the motion of the system as well as assess the engineering design of the system.
13	Simulation of Recreational Waterslide		Single-body Dynamic Analysis	1. This lesson presents a real-life application, simulation of riding object on a recreational waterslide using *Motion*. 2. A brief discussion on an analytical calculation method and experimental validation of the method is offered. 3. Discussion on a few modeling considerations in using *Motion* is also provided.

Table 1-5 Examples Employed in This Book (Cont'd)

Lesson	Title	Motion Model	Problem Type	Things to Learn
14	Quarter suspension of a racecar		Multibody Kinematic Analysis	1. This lesson discusses kinematic analysis of a quarter racecar suspension, including the addition of a cam representing the road profile. 2. Important kinematic measures of suspension design, including shock travel and camber angle, are included in discussion.

Lesson 2: Animations and Basic Motion
—A Single Piston Engine Example

2.1 Overview of the Lesson

In this lesson we learn the basic operations in *MotionManager* window. For those who are familiar with *the Animation* capabilities of *SOLIDWORKS Motion*, this lesson may serve as a review. If you are not familiar with *Animation*, this lesson will help you learn *MotionManager* window, especially in creating motion animation using *Animation* and *Basic Motion* studies. The skills you gain from this lesson should help you continuously learn in later lessons.

In this lesson, we will use the single piston engine discussed in *Lesson 1* as an example to illustrate the use of capabilities in *MotionManager* window. More specifically, we will use *Animation Wizard* to create an animation including rotation, explode, and collapse of the engine assembly. Then, we will add a rotary motor to drive the propeller for a kinematic analysis. This is a short and simple lesson. It should be an easy and fun lesson to go through.

2.2 The Single Piston Engine Example

Physical Model

The engine example consists of four components: engine case, propeller, connecting rod, and piston, as shown in Figure 2-1. The propeller is driven by a rotary motor at an angular speed of 60 RPM, i.e., one revolution per second. No gravity is present and English units system (IPS—inch, pound, second) is assumed.

Figure 2-1 The Engine Example

SOLIDWORKS Part and Assembly Files

For this lesson, the parts and assemblies have been created for you in *SOLIDWORKS*. There are a total of 22 *SOLIDWORKS* model files, eighteen parts and four assemblies. You can download these files from the publisher's web site (www.sdcpublications.com).

We will start with *Lesson2.SLDASM*, in which the engine is properly assembled with one free degree of freedom. When the propeller is driven by the rotary motor, it will rotate. The crankshaft where the propeller is mounted will then drive the connecting rod, and the connecting rod will push the piston up and down within the piston sleeve. Note that the assembly file *Lesson2withResults.SLDASM* contains a complete simulation model with simulation results. You may want to open this file to preview the motion model.

The assembly *Lesson2.SLDASM* consists of three sub-assemblies (*case_asm, propeller_asm,* and *connectingrod_asm*) and one part (*piston*). The *case_asm* is fixed (ground body). The *propeller_asm* is assembled to *case_asm* using concentric and coincident mates, as shown in Figure 2-2a. The propeller is free to rotate along the *X*-direction (referring to the reference triad). The *connectingrod_asm* is assembled to the propeller (at the crankshaft) using concentric and coincident mates, as shown in Figure 2-2b. The connecting rod is free to rotate relative to the propeller (at the crankshaft) along the *X*-direction. Finally, the piston is assembled to the connecting rod (at pin) using a concentric mate, as shown in Figure 2-2c. The piston is also assembled to the case using another concentric mate. This mate restricts the piston movement along the *Y*-direction, which in turn restricts the top end of the connection rod to move vertically.

(a) Mates between Case and Propeller   (b) Mates between Propeller and Rod   (c) Mates between Piston and Rod, and Piston and Case

Figure 2-2 Assembly Mates Defined for the Engine Example

Motion Studies

Two motion studies will be carried out. We will first create an animation for rotation, explode, and collapse of the engine assembly using *Animation Wizard*, where *Animation* will be chosen for the motion study. Note that an explode view has been created for this example. Then, we will add a rotary motor to drive the propeller for a kinematic analysis using *Basic Motion* study.

2.3   Using *SOLIDWORKS Motion*

Start *SOLIDWORKS* and open assembly file *Lesson2.SLDASM*. First, we take a look at the explode view by selecting the root node of the assembly (*Lesson2*), and press the right mouse button. In the menu appearing (Figure 2-3) choose *Explode*. You should see the assembly in an explode view similar to that of Figure 2-1. The animation will include exploding and collapsing the assembly. Right click the root assembly and choose *Collapse* to collapse the assembly.

Click the *Motion Study* tab (with the default name *Motion Study 1*) at the bottom of the graphics area to bring up the *MotionManager* window.

Creating Animation

We will create an animation that includes rotation, explode and collapse of the assembly using the *Animation Wizard. Animation Wizard* allows you to quickly create an animation.

Rotate Animation

Click the *Animation Wizard* button 🎬 from the *Motion* toolbar to bring up the *Animation Wizard* window (Figure 2-4a). Click *Rotate model* (default) and then *Next*.

The *Rotate model* option creates an animation by rotating the assembly around the *X-*, *Y-* or *Z*-axis. We will rotate the assembly along the *Y*-axis. Select *Y-axis*, enter *1* for *Number of rotations*, and click *Clockwise* for rotation direction (Figure 2-4b). Click *Next*. Enter *Duration (seconds): 5* (see Figure 2-4c), and *Start Time (seconds): 0* (default), then click *Finish*.

Note that as soon as you click *Finish*, a five-second animation timeline appears in the timeline area, as shown in Figure 2-5. The *Orientation and Camera Views* feature receives key points to mark the changes in the orientations as the assembly rotates.

Choose *Animation* (default) for the study type (right above the *MotionManager* tree, as circled in Figure 2-5), and click *Play From Start* ▮▶ from the *Motion* toolbar to play the animation from the beginning. The engine assembly should make a complete turn along the *Y*-axis. Click the *Stop* button ■ and drag the slider (right next to the *Stop* button) back to the initial position.

Figure 2-3

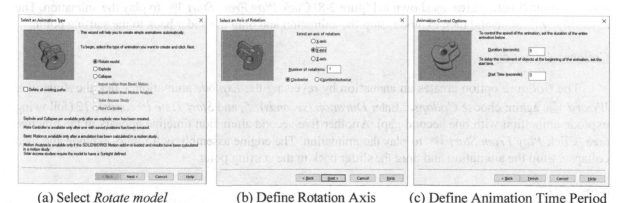

| (a) Select *Rotate model* | (b) Define Rotation Axis | (c) Define Animation Time Period |

Figure 2-4 The Animation Wizards

Explode Animation

The *Explode* option creates an animation by converting the explode view created in assembly into an animation in *Motion*. The sequence and the distances of the individual explode steps are used to create the animation. We will use the pre-defined explode view to create the animation.

Figure 2-5 The Timeline Area

Go back to *SOLIDWORKS* by clicking *Model* tab at the bottom of the graphics area. Select the *ConfigurationManager* above the *SOLIDWORKS* browser. Expand the *Default [Lesson2]* node, right click *ExpView1* and select *Explode*. You should see the explode view of the assembly appearing in the graphics area. Expand *ExpView1* to see the three explode steps, as shown in Figure 2-6. Click each step to see how they are defined.

Figure 2-6 *ConfigurationManager*

Now, we go back to *MotionManager* window by clicking *Motion Study 1* at the bottom of the graphics area.

Click the *Animation Wizard* button 📷, choose *Explode* (see Figure 2-7a) and then *Next*. Enter *Duration (seconds): 5*, (see Figure 2-7b) and *Start Time (seconds): 6* (that is, following rotation animation with one second gap), then click *Finish*. Note that as soon as you click *Finish*, the explode animation is appended to the rotate animation with one second gap. An additional five-second animation timeline appears in the timeline area, as shown in Figure 2-8. Click *Play From Start* ▶ to play the animation. The assembly will rotate and then explode. Stop the animation and drag the slider back to the starting point.

Collapse Animation

The *Collapse* option creates an animation by reversing the *Explode* animation. Click the *Animation Wizard* 📷 again; choose *Collapse*. Enter *Duration (seconds): 5*, and *Start Time (seconds): 12* (following explode animation with one second gap). Another five-second animation timeline appears in the timeline area. Click *Play From Start* ▶ to play the animation. The engine assembly will rotate, explode and then collapse. Stop the animation and drag the slider back to the starting point.

Save as AVI

Click *Save Animation* button 📷 from the *Motion* toolbar, and use the default names and settings. Click *Save* in the *Save Animation to File* window (Figure 2-9). Click *OK* in the *Video Compression* dialog box (Figure 2-10). In order to save the video, the animation will automatically play from the start and then record the animation. The time bar (vertical line) will run through the entire seventeen second animation period in the timeline area, and an *AVI* file (*Lesson2.avi*) will be created in the *Lesson 2* folder. You may use any *AVI* player, for example *Window Media Player*, to play the *AVI* video.

Figure 2-8 Added Timeline for
Explode Animation

(a) Select *Explode*   (b) Define Animation Time Period

Figure 2-7 The Animation Wizards

Figure 2-10

Figure 2-9

Rename the Motion Study

Change the name of the motion study from *Motion Study 1* (default) to *Animation* by double clicking the *Motion Study* tab, and enter *Animation*. Save your model before moving to the next motion study.

Creating a Motion Study using Basic Motion

We will create the next motion study by adding a rotary motor to the propeller, and use *Basic Motion* to carry out the motion study.

Create a new motion study by right clicking the *Motion Study* tab (just renamed *Animation*), and choose *Create New Motion Study*, as shown in Figure 2-11. Rename the study *Basic Motion*. You should see a fresh timeline area.

Choose *Basic Motion* from the motion study selection (right above the *MotionManager* tree, see Figure 2-12). Click the *Motor* button from the *Motion* toolbar to bring up the *Motor* window (Figure 2-13). Choose *Rotary Motor* (default).

Figure 2-11

Move the mouse pointer to the graphics area, and pick a circular arc of a solid feature of the propeller to define the rotational direction of the rotary motor; e.g., the circular arc on the drive washer of the propeller, as shown in Figure 2-14. A circular arrow appears indicating the rotational direction of the rotary motor.

Figure 2-12

Figure 2-13

Pick this Arc

Figure 2-14 Pick an Arc

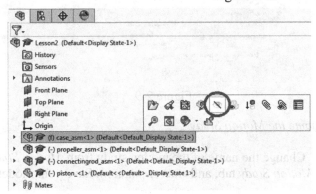

Figure 2-15 Hide Engine Case

Figure 2-16 Animation without the Case

A counterclockwise direction is desired. You may change the direction by clicking the direction button 🔁 under *Component/Direction*. Choose *Constant speed* and enter *60* RPM for speed. Click the checkmark ✅ on top of the dialog box to accept the motor definition. You should see a *RotaryMotor1* added to the *MotionManager* tree.

Click *Calculate* button 🖩 from the *Motion* toolbar to simulate the motion. A default five second simulation will be carried out. The propeller should make five turns (recall we entered 60 RPM for the motor earlier), and a five-second animation timeline will be created in the timeline area.

You may want to hide the *case_asm* to see how the connecting rod and piston move. Click *case_asm* in the *MotionManager* tree and select *Hide* 🖊 (circled in Figure 2-15). The engine case will be hidden from the graphics area. Play the animation again. You should now see the motion of connecting rod and piston like that of Figure 2-16.

Instead of repeating the same motion five times in five seconds, you may want to reduce the overall simulation period to one second and refine the time frame to create a smoother animation. Steps are described below.

Click the right key point (the end time key) of the key frame referring to the *Lesson2* node in the timeline area. You should see key point properties appear in the call out window, as shown in Figure 2-17. Drag the end time key to the one-second mark. Note the changes of all time frames in the timeline area. You may want to click the *Zoom In* button (at the right lower corner) to expand the timeline area. Calculate and play the animation. Now the propeller should rotate only once.

Figure 2-17

Click the *Motion Study Properties* button ⚙ from the *Motion* toolbar; the *Motion Study Properties* dialog box appears (see Figure 2-18). Change *Frames per second* to *100* under *Basic Motion*, and then click the checkmark ✅ on top to accept the change. Calculate and animate the motion again. The animation is smoother due to the increased frames per second.

Change the playback mode to loop 🔁 and play the animation continuously. The animation looks good and it is fun to see objects moving, isn't it? Save the model and move to the next lesson.

PropertyManager

Motion Study Properties

Animation ⌄

Basic Motion ⌃

Frames per second:

100

Geometry Accuracy:

Low — High

3D Contact Resolution:

Low — High

Motion Analysis ⌄

General Options ⌄

Figure 2-18 The *Motion Study Properties* Dialog Box

Exercises:

1. Open the propeller assembly and create an explode view like that of Figure E2-1. Use the *Animation Wizard* to create the following animations.

    (i)   Explode for 5 seconds, start time: 0 second.

    (ii) Rotate along Y-axis for 5 seconds, start time: 6 second.

    (iii) Rotate along X-axis for 5 seconds, start time: 12 second.

    (iv) Collapse for 5 seconds, start time: 18 second.

2. Use the animation created in Problem 1 with the following changes:

    (i)   Change crankshaft to red color for the entire animation. Hint: expand crankshaft node in the *MotionManager* tree, right click to choose *Appearance*, and select a red color.

    (ii) Change spinner to blue color, and only activate the change for the first five seconds of the animation. Hint: drag key points.

    (iii) Right click the *Orientation and Camera Views* node in the *MotionManager* tree and select *Disable View Key Creation* (see Figure E2-2). Choose the *Front* view, and play the animation. What is the difference you see in the animation? Check *SOLIDWORKS Help* to find out more about the *Orientation and Camera Views*.

Figure E2-2

Figure E2-1 Propeller in Explode View

Notes:

Lesson 3: A Ball Throwing Example

3.1 Overview of the Lesson

The lesson provides you with a quick run-through on *Motion Study* capabilities of *SOLIDWORKS Motion* using a simple ball-throwing example. This example simulates a ball thrown with an initial velocity at a given elevation. Due to gravity, the ball will travel following a parabolic trajectory and bounce back a few times when it hits the ground, as depicted in Figure 3-1. In this lesson, you will learn how to create a motion model to simulate the ball motion, run a simulation, and animate the ball motion. Simulation results obtained from *Motion* can be verified using particle dynamics theory that you learned in *Physics*. We will review the equations of motion, calculate the position and velocity of the ball, and compare our calculations with results obtained from *Motion*. Validating results obtained from computer simulations is extremely important. Note that *Motion* is not foolproof. In fact, several software bugs have been identified in *Motion 2024 SP1.0* (many are carried over from earlier versions). The software bug found in this example involves restitution coefficient of solid body contact resulting in erroneous results. There is no doubt that it requires readers to have a certain level of experience and expertise to master the software. Before you arrive at that level, it will be indispensable for you to verify the simulation results whenever possible. Verifying simulation results will increase your confidence in using the software and prevent you from being occasionally fooled by the erroneous simulations produced by the software. Often the erroneous results are due to modeling errors, and sometimes caused by software bugs.

3.2 The Ball Throwing Example

Physical Model

The physical model of the ball example is simple. The ball is made of *Cast Alloy Steel* with a radius of *10* in. The units system employed for this example is *IPS* (inch, pound, second). The gravitational acceleration is *386* in/sec². Note that you may check or change the units system by choosing from the pull-down menu

Tools → Options

Figure 3-1 The Ball Throwing Example

and choose the *Document Properties* tab in the *Document Properties-Units* dialog box. Click the *Units* node, and then pick the units system you prefer, as shown in Figure 3-2. The *IPS* should have been selected.

The ball and ground are assumed rigid. The ball will bounce back when it hits the ground. A coefficient of restitution $C_R = 0.75$ is specified to determine the bounce velocity when the impact occurs. For this example, $C_R = V_f/V_i$, where V_i and V_f are the velocities of the ball before and after the contact.

That is, the bounce velocity will be 75% of the incoming velocity in magnitude, and certainly, in the opposite direction. Note that in order for *Motion* to capture the moment when the ball hits the ground, we will define a solid body contact between the ball and the ground, and use the true geometry of the parts for a finer interference calculation during the simulation.

SOLIDWORKS Parts and Assembly

For this lesson, parts and assembly have been created for you in *SOLIDWORKS*. There are four files created: *ball.SLDPRT*, *ground.SLDPRT*, *Lesson3.SLDASM*, and *Lesson3withResults.SLDASM*. You can find these files at the publisher's web site (www.sdcpublications.com). We will start with *Lesson3.SLDASM*, in which the ball is fully assembled to the ground, i.e., no movement is allowed. In addition, the assembly file *Lesson3withResults.SLDASM* consists of a complete simulation model with simulation results, in which some of the assembly mates were suppressed in order to provide adequate degrees of freedom for the ball to move. You may want to open this file to see how the ball is supposed to move.

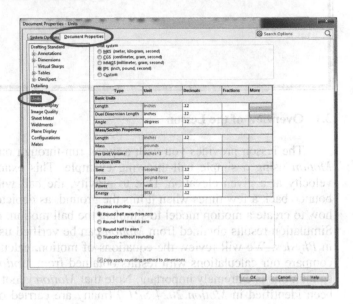

Figure 3-2 Checking Units System

Since gravity is defined in the negative *Y*-direction of the global coordinate system, all parts and assembly are created for a motion simulation that complies with the gravity setting.

The assembly *Lesson3.SLDASM* consists of two parts: the ball (*ball.SLDPRT*) and the ground (*ground.SLDPRT*). The ball is fully assembled to the ground by three assembly mates of three respective pairs of mating planes. They are *Front* (ball)/*Front* (ground), *Right* (ball)/*Right* (ground), and *Top* (ball)/*Top* (ground), as shown in Figure 3-3. The distance between the reference planes *Top* (ball) and *Top* (ground) is *100* in., which defines the initial position of the ball, as shown in Figure 3-4. The radius of the ball is *10* in., and the ground is modeled as a $30 \times 500 \times 0.04$ in.^{3} rectangular block.

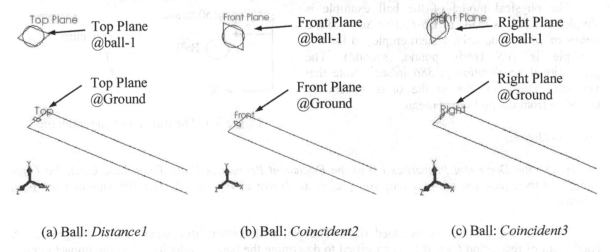

(a) Ball: *Distance1*   (b) Ball: *Coincident2*   (c) Ball: *Coincident3*

Figure 3-3 Assembly Mates Defined for the Example

Note that the *Y*-axis of the global coordinate system (or *Reference Triad* located at the lower left corner of the *SOLIDWORKS* graphics area, as shown in Figure 3-4) is pointing upward, which is consistent with the direction of the gravity but in the opposite direction.

100 in.

Ball

Groun

Global coordinate system
(or Reference Triad)

Motion Model

In this example, the ball will be the only movable body. Two assembly mates, *Distance1* and *Coincident3*, as shown in Figures 3-3a and 3-3c, will be suppressed to allow the ball to move on the *X-Y* plane. As mentioned earlier, the ball will be thrown with an initial velocity of $V_{0_x} = 150$ in/sec.

Figure 3-4 The *SOLIDWORKS* Assembly

A gravitational acceleration -386 in/sec^2 is defined in the *Y*-direction of the global coordinate system. The ball will reveal a parabolic trajectory due to gravity. A solid body contact will be added to characterize the contact between the ball and the ground. As discussed earlier, a coefficient of restitution $C_R = 0.75$ will be specified to determine the force that acts on the ball when the contact takes place. In this example, no additional friction is included.

3.3 Using *SOLIDWORKS Motion*

Start *SOLIDWORKS* and open assembly file *Lesson3.SLDASM*. Click the *Motion Study* tab at the bottom of the graphics area (*Motion Study 1*) to bring up the *MotionManager* window.

The *MotionManager* tree (or *Motion* browser) provides you with a graphical, hierarchical view of the motion model and allows you to edit existing motion entities conveniently by using the right click menu.

Switching back and forth between *Motion* and *SOLIDWORKS* assembly mode is straightforward. All you have to do is click the *Model* tab (to go back to *SOLIDWORKS*) and *Motion Study* tab (to *Motion*) when needed. As discussed in *Lesson 1*, the *Motion* toolbar, located at the top of the timeline area, provides settings, simulation and post-processing features for *Motion*. The *Play* button ▶ is especially handy when you finish running a simulation and are ready to animate the motion. Click some of the buttons and try to get familiar with their functions.

Before we start, we will suppress two assembly mates to allow the ball to move on the *X-Y* plane. Move the mouse pointer to *SOLIDWORKS* browser, and expand the *Mates* node by clicking the small ⊞ button in front of it. You should see three assembly mates: *Distance1(Ground<1>,ball<1>)*, *Coincident2(Ground<1>,ball<1>)*, and *Coincident3(Ground<1>,ball<1>)* listed in the browser, as shown in Figure 3-5.

Pick *Distance1(Ground<1>,ball<1>)*, and choose *Suppress* (circled in Figure 3-5). Repeat the same for *Coincident3(Ground<1>,ball<1>)*. Both mates will become inactive, and the ball is free to move on the *XY* plane. At this point, there is one ground body and one moving body (ball) in the motion model.

Please note that sometimes the mates suppressed in the *SOLIDWORKS* browser may be still active in the *MotionManager* tree. It is a good idea to expand the *Mates* node in both the *SOLIDWORKS* browser and *MotionManager* tree, and make sure the suppressed mates are grayed out (especially in *MotionManager* tree), as shown in Figure 3-6.

If you click the root node, *Lesson3 (Default<Display State-1>)* in the browser, you should see a rectangular box appears in the graphics area, which is simply the bounding box for the assembly (see Figure 3-7).

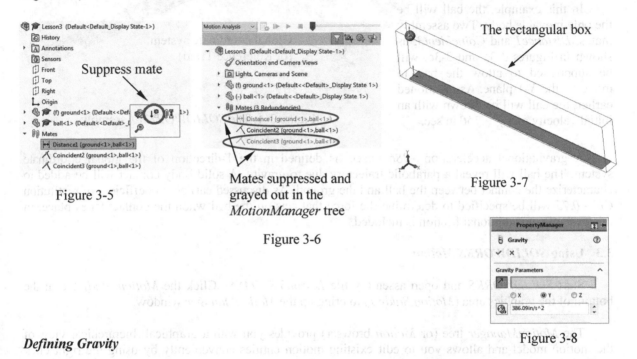

The rectangular box

Suppress mate

Figure 3-5

Mates suppressed and grayed out in the *MotionManager* tree

Figure 3-6

Figure 3-7

Figure 3-8

Defining Gravity

Click the *Gravity* button 🖥 from the *Motion* toolbar to bring up the *Gravity* dialog box. Choose *Y* and keep the g-value (386.09in/s^2) in the dialog box, as shown in Figure 3-8. In the graphics area, an arrow appears at the right lower corner ⬇, pointing downward indicating the direction of the gravity.

Click the checkmark ✅ on top of the dialog box to accept the gravity definition. A gravity node (name: *Gravity*) should appear in the *MotionManager* tree.

Defining Initial Velocity

We will add an initial velocity $V_{0_x} = 150$ in/sec to the ball. Click *ball* from the *MotionManager* tree, press the right mouse button, and choose *Initial Velocity* (see Figure 3-9a); the *Initial Velocity* dialog box will appear (Figure 3-9b).

From the graphics area, pick an edge on the ground part that is parallel to the *X*-axis like that in Figure 3-9c. Enter *150in/s*, as shown in Figure 3-9b. Click the checkmark ✅ to accept the initial velocity.

Running Simulation

Choose *Motion Analysis* for the study (see Figure 3-10). Click the *Motion Study Properties* button ⚙ from the *Motion* toolbar. In the *Motion Study Properties* dialog box (Figure 3-11), enter *100* for *Frames per second*, and click the checkmark on top. Drag the end time key to the three second mark (see Figure 3-12) in the timeline area to define the simulation duration. Click the *Zoom In* button to zoom in the timeline area if needed.

(a) Right Clicking the *ball* Node

(b) Entering Parameters

(c) Choosing Direction

Pick this edge

Direction arrow

Figure 3-9 Defining Initial Velocity

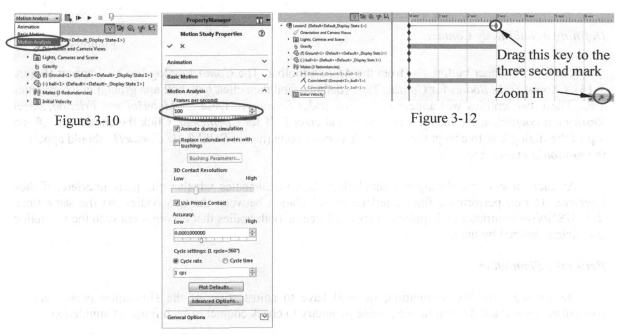

Figure 3-10

Figure 3-11

Figure 3-12

Drag this key to the three second mark

Zoom in

Click the *Calculate* button ▦ from the *Motion* toolbar to simulate the motion. A three second motion simulation will be carried out. The ball will be dropped from 100 in. above the ground with an initial horizontal velocity of 150 in/sec. After a few seconds, the ball starts moving. The ball falls through the ground and continues falling, as shown in Figure 3-13, which is not desirable. Note that the trace path that indicates the trace of the center of the ball is turned on in Figure 3-13. We will learn how to do this later. For the time being we will add a solid body contact between the ball and the ground in order to make the ball bounce back when it hits the ground. The contact will create a force to prevent the ball from penetrating the ground. This solid body contact will only be activated if the ball comes into contact with the ground.

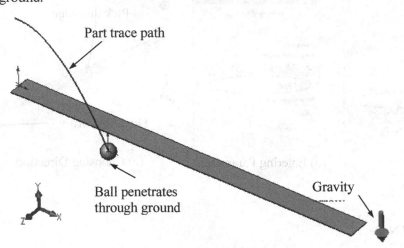

Part trace path

Ball penetrates
through ground

Gravity

Figure 3-13 Animation: Ball Penetrating Through the Ground

Figure 3-14

Defining a Solid Body Contact

Click the *Contact* button ⬚ from the *Motion* toolbar. The *Contact* dialog box will appear (Figure 3-14). Choose *Solid Bodies* for *Contact Type* (default), and then click both ball and ground in the graphics area. These two entities will appear in the box under *Selections*. Deselect *Material* and *Friction*, select *Restitution coefficient* for *Elastic Properties*, and enter *0.75* for *Coefficient*. Click the checkmark ✔ on top of the dialog box to accept the contact. A contact node (name: *Solid Body Contact1*) should appear in the *MotionManager* tree.

At each time frame during the simulation, *Motion* calculates whether the parts interfere. If they interfere, *Motion* performs a finer interference calculation between the two bodies. At the same time, *ADAMS/Solver* computes and applies an impact force on both bodies that is consistent with the restitution coefficient entered by the user.

Rerun the Simulation

Before we rerun the simulation, we will have to adjust some of the simulation parameters. In particular, we will ask *Motion* to use precise geometry to check contact in each frame of simulation.

Click the *Motion Study Properties* button ⚙ from the *Motion* toolbar. In the *Motion Study Properties* dialog box (Figure 3-15), choose *Use Precise Contact*. Click the checkmark to accept the contact constraint. Also, bring the ball to its initial position by dragging the time line (vertical line as shown in Figure 3-16) to 0 sec in the timeline area, or drag the slider right next to the *Stop* button in the *Motion* toolbar all the way to the left (beginning).

Click the *Calculate* button 🖩 again to rerun the simulation. After a few seconds, the ball starts moving. As shown in Figure 3-17 (no trace path shown yet) the ball will hit the ground and bounce back a few times before the simulation ends. The ball did not fall through the ground this time due to the addition of the solid body contact. We will learn how to create a trace path next. Note that if you are using *Motion 2024 SP1.0*, you will see the trace of the ball like that of Figure 3-17, which is, unfortunately, incorrect. We will discuss this shortly. Furthermore, analytical calculations and verification of the simulation results are included in Section 3.4.

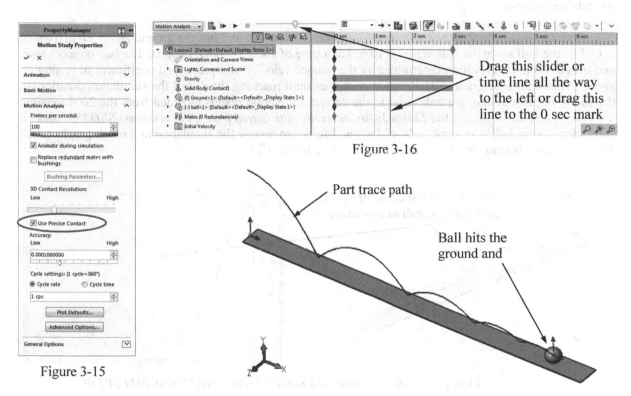

Figure 3-16

Figure 3-15

Figure 3-17 Animation: Ball Bouncing Back a Few Times

Displaying Simulation Results

Motion allows you to graphically display the path that a chosen point on the moving part follows. This is called a trace path. Note that the trace path of the ball was displayed in both Figures 3-13 and 3-17. We learn how to create a part trace path next.

Click the *Results and Plots* button 📊 from the *Motion* toolbar. In the *Results* dialog box (Figure 3-18), choose *Displacement/Velocity/Acceleration* from the first drop-down field. Then select *Trace Path* for the sub-category. Pick the origin point of the ball from the graphics area, and click the checkmark. A *Results* (and *Plot1<Trace Path1>*) node will be added to the *MotionManager* tree. Click the *Calculate* button again. When the ball starts moving, you should see a trace path following the ball like that of Figure 3-17.

Figure 3-18

Next, we will create a graph for the *Y*-position of the ball. Right click *ball* in the *MotionManager* tree, and choose *Create Motion Plot*. In the *Results* dialog box, choose *Displacement/Velocity/Acceleration*, select *Center of Mass Position*, and then *Y Component*. Click the checkmark.

A graph of the center of mass of the ball in the *Y*-direction will appear, similar to that of Figure 3-19 (which is incorrect) if you are running *Motion 2024 SP1.0*. The software bug was reported to the SOLIDWORKS company on 12/17/2009 (SR 1-1461525961). Figure 3-20 shows correct results obtained from an earlier version of *Motion* (more precisely *COSMOSMotion* 2008). Note that you may adjust properties of the graph, for instance the axis scales, following steps similar to those of Microsoft® Excel spreadsheet graphs.

Both graphs show that the ball was thrown from *Y = 100* in. and hits the ground at *Y = 10* in. (which is CM of the ball since the ball radius is 10 in.) at a time of about *t = 0.64* seconds. The ball bounces back and moves up to an elevation determined by the bounce velocity of a prescribed coefficient of restitution and gravitational acceleration. The motion continues until reaching the end of the simulation duration. In the following discussion, we only refer to the correct simulation results recorded from the 2008 version. You may open the assembly file *Lesson3withCorrectResults (saved from 2008 versions).SLDASM* under a subfolder *Lesson3withCorrectResults* of *Lesson 3* folder to review the simulation and result graphs (click *Play from Start* button ▯▶, do not click the *Calculate* button 🕮).

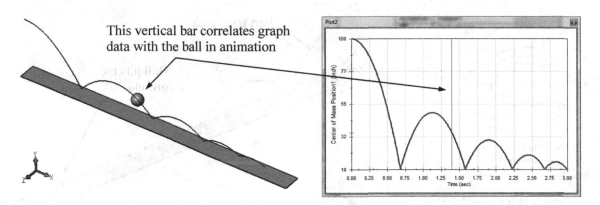

Figure 3-19 Result Graph and Motion Animation (*Motion 2024 SP1.0*)

Figure 3-20 Correct Result Graph and Motion Animation (Recorded from 2008 Version)

Note that you may click any location in the graph to bring up a fine red vertical line that correlates the graph with the position of the ball in animation. As shown in the graph of Figure 3-20, at $t = 1.4$ seconds, the ball is roughly at $Y = 44$ in. The snapshot of the ball at that specific time and its Y-location in the graph is shown in the graphics area.

Add two more graphs for the ball, the X-velocity and Y-velocity. As shown in Figure 3-21, the X-velocity of the ball is a constant *150* in/sec, which is equal to the initial velocity. This is because we turned off all friction for the solid body contact earlier (see Figure 3-14); therefore, no energy loss due to contact (except for the energy loss caused by the less than 1.0 restitution coefficient). Figure 3-22 shows *the Y*-velocity of the ball. The ball moves with a linear velocity due to gravity. The Y-velocity is about -264 in/sec at $t = 0.68$ seconds right before the ball contacts the ground. You may see this data by moving the point close to the corner point of the curve and leaving the point for a short period. The velocity data will appear.

Figure 3-21 *X*-Velocity of the Ball CM   Figure 3-22 *Y*-Velocity of the Ball CM

You may also convert the graph data to a Microsoft® Excel spreadsheet by simply moving the pointer inside the graph and right click to choose *Export CSV*. Open the spreadsheet to see more detailed simulation data. From the spreadsheet (Figure 3-23), the Y-velocity right before and after the ball hits the ground is *-264* and *198* in/sec, respectively. The ratio of *198/264* is about *0.75*, which is the coefficient of restitution we defined earlier. Note that the spreadsheet results were obtained from an earlier version (*COSMOSMotion* 2008).

If you use *Motion 2024 SP1.0*, the results you obtained are incorrect. If you check the Y-velocity data before and after the ball hits the ground to calculate the restitution coefficient, it does not equal 0.75. In fact, the restitution coefficient is calculated as 173.2/262.5 = 0.66 (not 0.75) at time = 0.68 second, as shown in Figure 3-24. The filename of the spreadsheet is *Linear Velocity 1*, which can be found in the Lesson 3 example folder.

In addition, the coefficient is not even a constant at the other bounce points if you check more locations. For example, at time = 1.58 second, the coefficient is 124.6/173.1 = 0.72 (again, not 0.75), as shown in Figure 3-25. This result is obviously incorrect.

	CM Velocity - Y-ball-1.csv		
	A	B	C
111	0.648	-250.271	
112	0.654	-252.588	
113	0.66	-254.906	
114	0.666	-257.223	
115	0.672	-259.54	
116	0.678	-261.857	
117	0.684	-264.175	
118	0.684362	198.0581	
119	0.69	195.8807	
120	0.696	193.5633	
121	0.702	191.246	
122	0.708	188.9287	
123	0.714	186.6114	
124	0.72	184.2941	
125	0.726	181.9767	
126	0.732	179.6594	
127	0.738	177.3421	
128	0.744	175.0248	
129	0.75	172.7074	
130	0.756	170.3901	

Figure 3-23

You may save the animation as an *AVI* movie file, following steps described in *Lesson 2*.

Be sure to save the model before exiting from *Motion.*

64	0.61	-235.514
65	0.62	-239.375
66	0.63	-243.236
67	0.64	-247.097
68	0.65	-250.958
69	0.66	-254.818
70	0.67	-258.679
71	0.68	-262.54
72	0.6828867	173.2354
73	0.69	170.489
74	0.7	166.6282
75	0.71	162.7673
76	0.72	158.9064
77	0.73	155.0455
78	0.74	151.1846
79	0.75	147.3237
80	0.76	143.4628
81	0.77	139.602
82	0.78	135.7411

Average: -21.98548797   Count: 4

Figure 3-24

157	1.53	-153.825
158	1.54	-157.686
159	1.55	-161.547
160	1.56	-165.408
161	1.57	-169.269
162	1.58	-173.13
163	1.5815828	124.635
164	1.59	121.3853
165	1.6	117.5244
166	1.61	113.6635
167	1.62	109.8026
168	1.63	105.9417
169	1.64	102.0808
170	1.65	98.21994
171	1.66	94.35905
172	1.67	90.49817
173	1.68	86.63728
174	1.69	82.7764
175	1.7	78.91551

Average: -11.33329228   Count: 4

Figure 3-25

3.4   Result Verifications

In this section, we will verify analysis results obtained from *Motion* using particle dynamics theory we learned in *Physics*.

There are two assumptions that we have to make in order to apply the particle dynamics theory to this ball-throwing problem:

(i)   The ball is of a concentrated mass, and
(ii) No air friction is present.

Equation of Motion

It is well-known that the equations of motion that describe the position and velocity of the ball are, respectively,

$$P_x = P_{0_x} + V_{0_x} t \tag{3.1a}$$

$$P_y = P_{0_y} + V_{0_y} t - \frac{1}{2} g t^2 \tag{3.1b}$$

and

$$V_x = V_{0_x} \tag{3.2a}$$

$$V_y = V_{0_y} - g t \tag{3.2b}$$

where P_x and P_y are the X- and Y-positions of the ball, respectively; V_x and V_y are the X- and Y-velocities, respectively; P_{0x} and P_{0y} are the initial positions in the X- and Y-directions, respectively; V_{0x} and V_{0y} are the initial velocities in the X- and Y-directions, respectively; and g is the gravitational acceleration.

These equations can be implemented using, for example, Microsoft® Excel spreadsheet shown in Figure 3-26, for numerical solutions. In Figure 3-26, Columns B and C show the results of Eqs. 3.1a and 3.1b, respectively; with a time interval from *0* to *3* seconds and increment of *0.01* seconds. Columns D and E show the results of Eqs. 3.2a and 3.2b, respectively.

Note that when the ball hits the ground, we will have to reset the velocity to 75% of that at the prior time step and in the opposite direction.

Data in column C is graphed in Figure 3-27. Columns D and E are graphed in Figure 3-28. Comparing Figure 3-27 with Figure 3-20 and Figure 3-28 with Figures 3-21 and 3-22, the results obtained from theory and *Motion* should be very close (but not the case for *Motion 2024 SP1.0*), which means the dynamic model has been created properly in *Motion*. Note that the solution spreadsheet can be found at the publisher's website (filename: *lesson3.xls*).

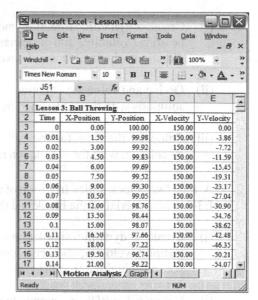

Figure 3-26 The Excel Spreadsheet

Figure 3-27 Graph of the *Y*-Position of the Ball Obtained from Spreadsheet Calculations

Figure 3-28 Graph of the *X*- and *Y*-Velocities of the Ball Obtained from Spreadsheet Calculations

Exercises:

1. Use the same motion model to conduct a simulation for a different scenario. This time the ball is thrown at an initial velocity of $V_{0_x} = 100$ in/sec and $V_{0_y} = 50$ in/sec from an elevation of *150* in., as shown in Figure E3-1.

    (i) Create a dynamic simulation model using *Motion* to simulate the trajectory of the ball. Report position, velocity, and acceleration of the ball at *0.2* seconds in both vertical and horizontal directions obtained from *Motion*.

    (ii) Derive and solve the equations that describe the position and velocity of the ball. Compare your solutions with those obtained from *Motion*.

    (iii) Calculate the time for the ball to reach the ground and the distance it travels. Compare your calculation with the simulation results obtained from *Motion*.

2. A *1"×1"×1"* block slides from top of a *45°* slope (due to gravity) without friction, as shown in Figure E3-2. The material of the block and the slope is *AL2014*.

    (i) Create a dynamic simulation model using *Motion* to analyze motion of the block. Report position, velocity, and acceleration of the block in both vertical and horizontal directions at *0.2* seconds obtained from *Motion*.

    (ii) Create a *LimitDistance* mate to stop the block when its front lower edge reaches the end of the slope. You may want to review *SOLIDWORKS* help menu or preview *Lesson 5* to learn more about the *LimitDistance* mate.

    (iii) Derive and solve the equation of motion for the system. Compare your solutions with those obtained from *Motion*.

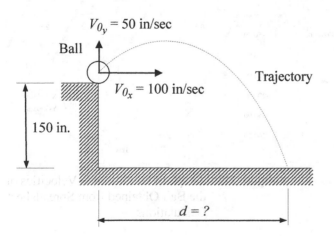

Figure E3-1 The Ball Throwing Problem

Figure E3-2 The Block Sliding Problem

Lesson 4: A Simple Pendulum

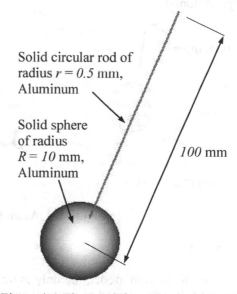

4.1 Overview of the Lesson

In this lesson, we create a simple pendulum motion model using *SOLIDWORKS Motion*. The pendulum will be released from a position slightly off from the vertical direction. Once released, the pendulum rotates freely due to gravity. In this lesson, we will learn how to create a pendulum motion model, run a dynamic analysis, and visualize the analysis results. The dynamic analysis results of the simple pendulum example can be verified using particle dynamics theory. Similar to *Lesson 3*, we will formulate the equation of motion; calculate the angular position, velocity, and acceleration of the pendulum; and compare our calculations with results obtained from *Motion*.

4.2 The Simple Pendulum Example

Physical Model

The physical model of the pendulum is composed of a sphere and a rod rigidly connected, as shown in Figure 4-1. The radius of the sphere is *10* mm. The length and radius of the thin rod are *90* mm and *0.5* mm, respectively. The top of the rod is connected to the wall with a revolute or pin joint (assembled using concentrate and coincident mates). This revolute joint allows the pendulum to rotate. More about the revolute joint (and other joint types) can be seen in Appendix A. Both rod and sphere are made of Aluminum. Note that from the *SOLIDWORKS* material library *Aluminum Alloy 2014* has been selected for both sphere and rod. The *MMGS* units system is chosen for this example (millimeter for length, Newton for force, and second for time). Note that in the *MMGS* units system, the gravitational acceleration is *9,806* mm/sec².

Solid circular rod of radius *r = 0.5* mm, Aluminum

Solid sphere of radius *R = 10* mm, Aluminum

100 mm

Figure 4-1 The Pendulum Physical Model

The pendulum will be released from an angular position of *10* degrees measured from the vertical position about the rotational axis of the revolute joint. The rotation angle is intentionally kept small so that the particle dynamics theory can be applied to verify the simulation result.

SOLIDWORKS Part and Assembly Files

In this lesson, *SOLIDWORKS* parts and assemblies of the pendulum example have been created for you. There are four model files created: *pendulum.SLDPRT*, *ground.SLDPRT*, *Lesson4.SLDASM*, and *Lesson4withResults.SLDASM*. You can find these files at the publisher's web site (www.sdcpublications.com). We will start with *Lesson4.SLDASM*, in which the pendulum is fully assembled to the ground. In addition, the assembly file *Lesson4withResults.SLDASM* consists of a complete simulation model with simulation results.

In the assembly models, there are three assembly mates, *Concentric1(ground<1>,pendulum<1>)*, *Coincident1(ground<1>, pendulum <1>)*, and *Angle1(pendulum <1>, Right Plane)*, as shown in Figure 4-2. Note that the first two mates restrict the pendulum to rotate about the *Z*-axis of the global coordinate system. The third mate, *Angle1*, keeps the pendulum at a negative *10* degree angle about the *Z*-axis. This mate is required to position the pendulum for an initial condition. You may want to use the *Front* view (one of *SOLIDWORKS* predefined views located at the top of the graphics area) to see how the pendulum is oriented. The pendulum is fully constrained; no movement is allowed. We will suppress the third mate *Angle1* before entering *Motion*.

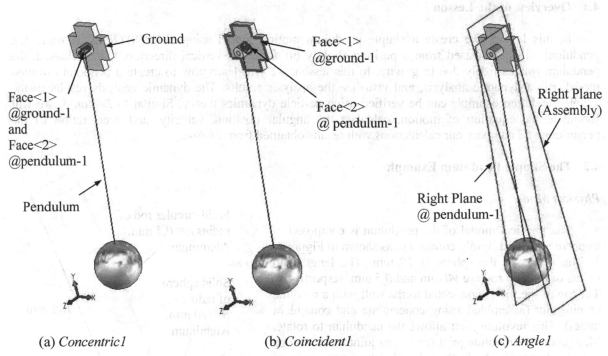

(a) *Concentric1*   (b) *Coincident1*   (c) *Angle1*

Figure 4-2 Assembly Mates Defined in *Lesson4.SLDASM*

Motion Model

In this motion model, the only movable body is the pendulum. After suppressing the mate *Angle1*, the pendulum is allowed to rotate about the *Z*-axis of the global coordinate system due to gravity and a small initial angle. Note that gravity is acting in the negative *Y*-direction. Friction between the pendulum and the ground is assumed to be zero. You may want to open the assembly file *Lesson4withResults.SLDASM* to play the animation before starting the lesson.

4.3 Using SOLIDWORKS Motion

Start *SOLIDWORKS* and open the assembly file *Lesson4.SLDASM*. First, suppress the third assembly mate *Angle1* by going over the following steps.

From the *SOLIDWORKS* browser, expand the *Mates* node, click *Angle1*, and choose *Suppress* ↓🖫. The mate *Angle1* will become inactive. Save the model.

Note that most steps in this lesson are similar to those of *Lesson 3*. You may see some repetitions. Hopefully, with some steps repeated here, you will not have to go back to *Lesson 3* for references.

Click the *Motion Study* tab (*Motion Study 1*) at the bottom of the graphics area to bring up the *MotionManager* window. Like *Lesson 3*, we will add gravity and create a 1.5-second simulation.

Defining Gravity

Click the *Gravity* button ⌀ from the *Motion* toolbar to bring up the *Gravity* dialog box. Choose *Y* and keep the g-value (9806.65mm/s^2) in the dialog box, as shown in Figure 4-3. In the graphics area, an arrow appears at the lower right corner ↓, pointing downward indicating the direction of the gravity. Click the checkmark ✔ on top of the dialog box to accept the gravity.

Figure 4-3

Running Simulation

Choose *Motion Analysis* for the study (see Figure 4-4). Click the *Motion Study Properties* button ⚙ from the *Motion* toolbar. In the *Motion Study Properties* dialog box (Figure 4-5), enter *300* for *Frames per second*, and click the checkmark. Drag the end time key to 1.5 second mark (see Figure 4-6) in the timeline area to define the simulation duration. Click the *Zoom In* button to zoom in the timeline area.

Figure 4-4

Figure 4-5

Click *Calculate* button 🖩 from the *Motion* toolbar to simulate the motion. A 1.5 second simulation will be carried out. You should see that the pendulum starts moving back and forth about the axis of the revolute joint (if not, check if mate *Angle1* is suppressed in *MotionManager* tree). We will graph the position, velocity, and acceleration of the pendulum next.

Drag this key to the 1.5 second mark

Zoom in

Figure 4-6

Displaying Simulation Results

The results of the angular position, velocity, and acceleration of the pendulum can be directly obtained by right clicking the moving part, *pendulum-1*, from the *MotionManager* tree.

Right-click *pendulum-1*, and choose *Create Motion Plot*. In the *Results* dialog box, choose *Displacement/Velocity/Acceleration*, select *Angular Displacement*, and then *Magnitude*. Click the checkmark.

A graph like that of Figure 4-7 should appear. From the graph, the pendulum rotates about the Z-axis between −10 and 10 degrees, as expected (since no friction is involved). Also, it takes about 0.6 seconds to complete a cycle. Note that you can export the graph data, for example, by right clicking the graph and choosing *Export CSV*. Open the spreadsheet and examine the data. From the spreadsheet, the time for the pendulum to swing back to its original position, i.e., −10 degrees, is 0.64 seconds, as shown in the spreadsheet of Figure 4-8. We will carry out calculations to verify these results later. Before we do that, we will graph the angular velocity and acceleration of the pendulum.

Figure 4-7 Rotation Angle of the Pendulum

Figure 4-8

Figure 4-9 *Z*-Angular Velocity of the Pendulum   Figure 4-10 *Z*-Angular Acceleration of the Pendulum

Repeat the same steps and in the *Results* dialog box, choose *Angular Velocity* and *Z Component*. Repeat the same steps and choose *Angular Acceleration* and *Z Component*. You should see graphs like those of Figures 4-9 and 4-10. Figure 4-9 shows that the angular velocity starts at 0, which is expected.

The angular velocity varies roughly between *–100* and *100* degrees/sec. Also, the angular acceleration varies roughly between *–1,000* and *1,000* degrees/sec². Are these results correct? We will carry out calculations to verify if these graphs are accurate.

Save the model.

4.4 Result Verifications

In this section, we will verify the analysis results obtained from *Motion* using particle dynamics theory.

There are four assumptions that we have to make in order to apply the particle dynamics theory to this simple pendulum problem:

(i) Mass of the rod is negligible (this is why the diameter of the pendulum rod is so small),
(ii) The sphere is of a concentrated mass,
(iii) Rotation angle is small (remember the initial conditions we defined?), and
(iv) No friction is present.

The pendulum model has been created to comply with these assumptions as much as possible. We expect that the particle dynamics theory will give us results close to those obtained from simulation. Two approaches will be presented to formulate the equations of motion for the pendulum: energy conservation and Newton's law.

Energy Conservation

Referring to Figure 4-11, the kinetic energy and potential energy of the pendulum can be written, respectively, as

$$T = \frac{1}{2} J \dot{\theta}^2 \qquad (4.1)$$

where J is the moment of inertia, i.e., $J = m\ell^2$ in this case; and

$$U = mg\ell \ (1 - \cos \theta) \qquad (4.2)$$

According to the energy conservation theory, the total mechanical energy, which is the sum of the kinetic energy and potential energy, is a constant with respect to time, i.e.,

$$\frac{d}{dt}(T + U) = 0 \qquad (4.3)$$

where t represents time. Hence

Figure 4-11 Particle Dynamics of Pendulum

$$\frac{d}{dt}\left(\frac{1}{2}m\ell^2 \dot{\theta}^2 + mg\ell(1 - \cos\theta)\right) = m\ell^2 \ddot{\theta} + mg\ell \sin\theta = 0 \qquad (4.4)$$

Therefore,

$$\ddot{\theta} + \frac{g}{\ell} \sin\theta = 0 \text{, and}$$

$$\ddot{\theta} + \frac{g}{\ell}\theta = 0 \qquad\qquad\qquad (4.5)$$

when $\theta \approx 0$.

Newton's Law

From the free-body diagram shown in Figure 4-12, the equilibrium equation of moment at the origin about the Z-axis (normal to the paper) can be written as:

$$\sum M = -mg\ell \sin\theta = J\ddot{\theta} = m\ell^2 \ddot{\theta} \qquad (4.6)$$

Hence

$$\ddot{\theta} + \frac{g}{\ell}\sin\theta = 0 \text{, and}$$

$$\ddot{\theta} + \frac{g}{\ell}\theta = 0 \qquad\qquad\qquad (4.7)$$

Figure 4-12 Free Body Diagram

when $\theta \approx 0$.

Note that the same equation of motion has been derived from two different approaches. The linear ordinary second-order differential equation can be solved analytically.

Solving the Differential Equation

It is well known that the solution of the differential equation is

$$\theta = A_1 \cos\omega_n t + A_2 \sin\omega_n t \qquad\qquad\qquad (4.8)$$

where $\omega_n = \sqrt{\frac{g}{\ell}}$, and A_1 and A_2 are constants to be determined by initial conditions. Note that

$\omega_n = \sqrt{\frac{g}{\ell}} = \sqrt{\frac{9806}{100}} = 9.903$ rad/sec, and the natural frequency of the system is $f_n = \omega_n/2\pi = 1.576$ Hz. The period for a complete cycle is $T = 1/f_n = 0.634$ seconds, which is close to that shown in all graphs (for example, Figure 4-7) as well as the spreadsheet data shown in Figure 4-8 generated using *Export CSV*.

The initial conditions for the pendulum are $\theta(0) = \theta_0 = -10$ degrees, and $\dot{\theta}(0) = 0$ degree/sec. Plugging the initial conditions into the solution, we have

$A_1 = \theta_0 = -10$ degrees, and

$A_2 = 0.$

Hence, the solutions are

$$\theta = \theta_0 \cos \omega_n t \qquad\qquad (4.9a)$$

$$\dot{\theta} = -\theta_0 \omega_n \sin \omega_n t \qquad\qquad (4.9b)$$

$$\ddot{\theta} = -\theta_0 \omega_n^2 \cos \omega_n t \qquad\qquad (4.9c)$$

The above equations represent angular position, velocity, and acceleration of the revolute joint, respectively. These equations can be implemented into, for example, *Excel* spreadsheet shown in Figure 4-13, for numerical solutions. Columns B, C, and D in the spreadsheet show the results of Eqs. 4.9a, b, and c, respectively, between *0* and *1.5* seconds with increments of *0.005* seconds. Data in these three columns are graphed in Figures 4-14, 4-15, and 4-16, respectively. Comparing Figures 4-14 to 4-16 with Figures 4-7, 4-9, and 4-10, the results obtained from theory and simulation are close. The motion model has been properly defined, and *Motion* gives us reliable results. Note that in the calculation, the angular position of the pendulum is set to zero when it aligns with the vertical axis; therefore, the pendulum swings between *–10* and *10* degrees.

However, even though graphs obtained from *Motion* and spreadsheet calculations are alike, these results are not identical. This is because the *Motion* model is not a perfect simple pendulum since mass of the rod is non-zero. If you reduce the diameter of the rod, the *Motion* results should approach those obtained through spreadsheet calculations.

Figure 4-13 The *Excel* Spreadsheet

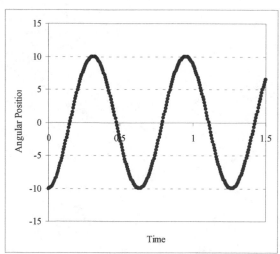

Figure 4-14 Angular Position from Theory

Figure 4-15 Angular Velocity from Theory

Figure 4-16 Angular Acceleration from Theory

Exercises:

1. Create a spring-damper-mass system, as shown in Figure E4-1, using *Motion*. Note that the unstretched spring length is *3* in. The radius of the ball is *0.5* in. and the material is *Cast Alloy Steel* (mass density: *0.2637* lb_m/in^3).

    (i) Find the spring length in the equilibrium condition using *Motion*.

    (ii) Solve the same problem using Newton's laws. Compare your results with those obtained from *Motion*.

2. If a force $p = 2$ lb_f is applied to the ball as shown in Figure E4-1, repeat both (i) and (ii) of Problem 1.

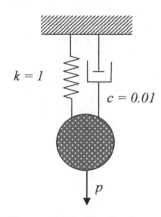

$k = 1$

$c = 0.01$

p

Figure E4-1 The Spring-
        Mass-Damper System

Notes:

Lesson 5: Friction of a Sliding Object

5.1 Overview of the Lesson

In this lesson, we study friction of a sliding object. There are two approaches to define friction for a sliding object in *SOLIDWORKS Motion*. They are contact friction and joint friction. You may create contact friction by defining a solid body contact between the objects; for example, a sliding block and the surface that the block slides on, as shown in Figure 5-1. In defining contact friction, you may specify friction by either selecting materials for the respective objects in contact, entering friction coefficients, or specifying elastic properties (such as restitution coefficient we learned in *Lesson 3*). You may also define joint friction as part of assembly mate between the sliding object and the surface it slides on by specifying friction coefficients.

In this lesson, we will learn both approaches using a simple example shown in Figure 5-1, in which a steel block of *1*in.×*1*in.×*1*in. is sliding along a *30⁰* slope face due to gravity. We will learn how to create the sliding motion model, run a motion analysis, and visualize the analysis results. We will learn how to add friction using both approaches. Part of the reason for going over this sliding example by simply specifying friction coefficients is that the analysis results of the example can be verified using particle dynamics theory we learned in *Physics*. Similar to *Lessons 3* and *4* we will formulate the equation of motion, solve the equation, and compare our calculations with results obtained from *Motion*.

5.2 The Sliding Object

Physical Model

Note that the *IPS* units system will be used for this example. The block slides along the *30⁰* slope face for a distance of *9* in. with friction specified by friction coefficients. We assume friction coefficients of $\mu = 0$ (no friction), $\mu = 0.01$, 0.02, 0.03, 0.05, 0.075, 0.1, 0.2, 0.3, 0.4, and 0.5.

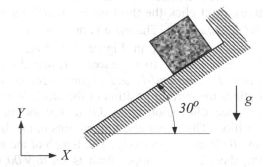

Figure 5-1 The Sliding Friction Example

SOLIDWORKS Part and Assembly Files

For this lesson, the part and assembly files have been created in *SOLIDWORKS*. There are five model files created: *block.SLDPRT*, *ground.SLDPRT*, *Lesson5.SLDASM*, *Lesson5ContactFrictionwithResults.SLDASM*, and *Lesson5JointFrictionwithResults.SLDASM*. Note that the assembly files *Lesson5ContactFrictionwithResults.SLDASM* and *Lesson5JointFrictionwithResults.SLDASM* are motion models that contain the complete simulation results obtained by defining friction through solid body contact and joint, respectively. You may open these two files to preview the motion of the sliding object before starting this lesson. We will start this lesson by opening the assembly *Lesson5.SLDASM*, in which the block is assembled to *ground* without any motion entities.

In the assembly models, there are three assembly mates, *Coincident1(ground<1>,block<1>)*, *Coincident3(ground<1>,block<1>)*, and *LimitDistance1(ground<1>,block<1>)*, as shown in Figure 5-2. The block is allowed to move along the slope face. You may drag the block in the graphics area; you should be able to move the block on the slope face, but not beyond the face. This is because the third mate, *LimitDistance1*, is defined to restrict the block to move between the lower and upper limits. Choose *Edit → Undo Move Component* from the pull-down menu to restore the block to its initial position. We will take a look at the assembly mate *LimitDistance1*.

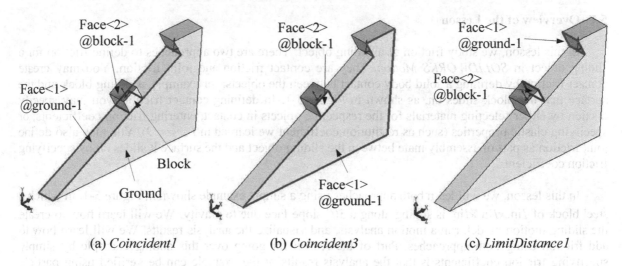

(a) *Coincident1*   (b) *Coincident3*   (c) *LimitDistance1*

Figure 5-2 Assembly Mates Defined in *Lesson5.SLDASM*

From *SOLIDWORKS* browser (or *FeatureManager* tree), right click the third mate, *LimitDistance1*, and choose *Edit Feature* 🔧. The mate is brought back for reviewing or editing, as shown in Figure 5-3. Note that the distance between the two faces (*Face<2>@block-1* and *Face<1>@ground-1*, see Figure 5-2c) is *3.00* in., which defines the current position of the block. We will change the distance dimension from *3.0* to *0.0* for the initial sliding position. The upper and lower limits of the distance are *9.00* and *0.00* in., respectively. The length of the slope face is *10* in.; therefore, the upper limit is set to *9.00* in., so that the block will stop when its front lower edge reaches the end of the slope face (since the block width is *1* in.). The *LimitDistance1* is an advanced mate in *SOLIDWORKS*.

Figure 5-3

Motion Models

As mentioned above, two approaches are supported in *Motion* for defining friction between bodies. As a result, we will create two motion models in this lesson. The first model employs solid body contact between the block and the slope face with different friction coefficients. We verify the results in Section 5.4 before discussing the second approach in Section 5.5, i.e., defining joint friction as part of the assembly mate. Gravitation acceleration will be turned on for this example, as shown in Figure 5-4. This is a simple motion model and yet offers a lot of insights for us to learn.

It is important to note that for the first motion model, in which we employ solid body contact for defining the friction, we will have to suppress *Coincident1* mate between the block and the slope face before entering *Motion*. This is to eliminate the effect of the coincident mate to the motion of the block, and allow the definition of solid body contact to determine the motion of the block.

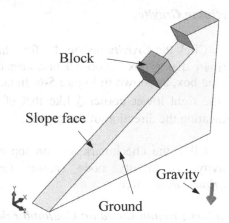

Figure 5-4 The Sliding Motion Model

In the second motion model, we will again suppress the *Coincident1* mate and define a local coincident mate (also called motion study specific mate) between the block and the slope face for individual motion simulations of different friction coefficients.

5.3 Using *SOLIDWORKS Motion*: Contact Friction

Start *SOLIDWORKS* and open the assembly file *Lesson5.SLDASM*.

Before creating any motion entities, we check the units system. Similar to *Lesson 3*, choose from the pull-down menu

Tools → Options

and choose the *Document Properties* tab in the *Document Properties-Units* dialog box, then click the *Units* node. The *IPS* should have been selected. Close the dialog box. In this units system, the gravity is *386* in/sec^2 in the negative *Y*-direction of the global coordinate system.

Defining Initial Sliding Position

In the *FeatureManager* tree, expand *Mates*, right click *LimitDistance1*, and choose *Edit Definition*. In the *LimitDistance1* dialog box (see Figure 5-5), enter *0* for distance dimension (circled in Figure 5-5). Click the checkmark ✔ on top of the dialog box to accept the change. The block is now moved to the starting position at the right end of the slope face.

Figure 5-5

Suppressing Coincident Mate: Coincident1

In the *FeatureManager* tree, right click *Coincident1* and choose *Suppress*. The mate should be suppressed and turned into gray color.

Entering Motion

Click the *Motion Study* tab (*Motion Study 1*) at the bottom of the graphics area to bring up the *MotionManager* window.

Defining Gravity

Click the *Gravity* button ⏻ from the *Motion* toolbar to bring up the *Gravity* dialog box. Choose *Y* and keep the g-value (386.09in/s^2) in the dialog box, as shown in Figure 5-6. In the graphics area, an arrow appears at the right lower corner ⬇ like that of Figure 5-4, pointing downward indicating the direction of the gravity.

Click the checkmark ✔ on top of the dialog box to accept the gravity. A gravity node (name: *Gravity*) should appear in the *MotionManager* tree.

Figure 5-6

Contact Friction Based on Coulomb Friction Method

In *SOLIDWORKS Motion*, friction defined between bodies in contact is based on Coulomb friction method, which is a bit different than joint friction to be discussed later in this lesson.

The Coulomb model of friction, or the Amonton-Coulomb model of friction, or Coulomb friction, refers to a simple model of dry friction that is fairly accurate empirically. Specifically, it is a model that helps predict the direction and magnitude of the friction force between two bodies with dry surfaces in contact.

The friction model we learned in high school *Physics* is the static friction of Coulomb model, in which we assume that there is no slipping between the two surfaces of contact. This can occur both when the two bodies are in rest relative to each other, and when the surfaces are rolling against each other. In this case, the maximum value of static friction that can be experienced through the surface of contact equals $\mu_s N$, where μ_s is the limiting coefficient of static friction (or simply static friction coefficient as we know) and N is the normal force. In the limiting case, the two surfaces are said to *just start slipping against each other*.

In fact, static friction is equal in magnitude and opposite in direction to the net external force acting along the plane of contact that is causing a tendency for the surfaces in contact to slip against each other. In other words, it acts to precisely cancel out the external force that would otherwise have caused slipping, thus producing no slipping.

In addition to static friction, Coulomb friction predicts kinetic friction, or dynamic friction, which occurs when there is slipping between the two surfaces of contact. There exists a constant μ_k, termed the coefficient of kinetic friction, dependent on the nature of the two surfaces in contact. The magnitude of kinetic friction force is $\mu_k N$, where N is the normal force between the bodies.

Note that kinetic friction force is opposite in direction to the direction of actual slipping. Its direction does not depend on the direction of other external forces.

Kinetic friction coefficient μ_k for a given pair of surfaces is typically somewhat less than static friction coefficient μ_s, indicating that once two bodies start slipping against each other, it is easier for them to continue slipping.

Contact Friction Supported in Motion

There are three methods provided in *Motion* for you to define contact friction. They are material, friction coefficients, and elastic properties. You may choose materials of the respective bodies in contact, specify friction coefficients for either static or dynamic friction, or define elastic properties through either impact or restitution coefficient. Please recall that we learned to use the restitution coefficient to determine the ball bouncing velocity in *Lesson 3*.

In this lesson, we learn to define friction in solid body contact, which is again based on Coulomb friction method.

The Coulomb friction method implemented in *Motion* is explained in Figure 5-7a, in which Coulomb friction forces are calculated based on a curve determined by four friction parameters, *static friction velocity* v_s, *dynamic friction velocity* v_k, *static friction coefficient* μ_s, and *dynamic friction coefficient* μ_k.

The static friction coefficient is the constant used to calculate the force necessary to overcome friction when a body is at rest. The dynamic friction coefficient is the constant used to calculate friction forces once the body is no longer at rest. In reality, the static friction velocity is zero, but in *Motion*, a non-zero value must be specified to avoid a singularity at the origin. More specifically, when a part is in transition from a negative to positive velocity, and when the velocity is zero, the force magnitude cannot instantaneously transition from a positive to negative value. Therefore, the curve in Figure 5-7a is how *Motion* resolves this issue: the user specifies static and dynamic velocities (see Figure 5-7b) where the friction coefficients are used. From there, *Motion* fits a smooth curve to solve for the friction force.

Note that when you choose material to define friction contact, *Motion* calculates friction by using the prescribed friction parameters of the materials selected for the respective bodies in contact and following the Coulomb friction model illustrated in Figure 5-7a. For example, the default friction parameters for *dry steel* in contact with *dry steel* are:

Static Friction Velocity: $v_s = 0.1$ mm/s
Dynamic Friction Velocity: $v_k = 10$ mm/s
Static Friction Coefficient: $\mu_s = 0.30$
Dynamic Friction Coefficient: $\mu_k = 0.25$

(a)

(b)

Figure 5-7 Illustration of Coulomb Friction Model Defined in *SOLIDWORKS Motion*

(b)

(a)

Figure 5-8 The Coulomb Friction for the Block Sliding Example

Instead of selecting material, the same friction model can be defined by selecting friction and entering the parameters using the *Contact* dialog box, as shown in Figure 5-7b.

With the basic understanding in Coulomb friction model, we are ready to define the friction for the block. We assumed a constant friction coefficient $\mu_k = 0.1$ that is activated from the beginning by entering $v_k = 0$. The Coulomb friction model is shown in Figure 5-8a, which can be defined by selecting friction and entering the friction parameters $v_k = 0$ and $\mu_k = 0.1$ using the *Contact* dialog box, as shown in Figure 5-8b.

Defining Solid Body Contact

Click the *Contact* button ⬚ from the *Motion* toolbar. The *Contact* dialog box will appear (Figure 5-9). Choose *Solid Bodies* for *Contact Type* (default), and then click both the block and the ground from the graphics area. These two entities will appear in the box under *Selections*. Deselect *Material*. Select *Friction*, enter *0* for v_k, and *0.1* for μ_k. Deselect *Static friction*. Turn off *Elastic Properties* by selecting *Restitution coefficient*, and enter *0* for *Coefficient*. Click the checkmark ✔ on top of the dialog box to accept the contact constraint. A solid body contact (name: *Solid Body Contact1*) should appear in the *MotionManager* tree.

Note that we set the restitution coefficient to be *0* in order to eliminate the possibility of the block bouncing up and down along the slope face, although physically this should not happen.

Figure 5-9

Defining and Running Simulation

Choose *Motion Analysis* for study. Click the *Motion Study Properties* button ⚙ from the *Motion* toolbar. In the *Motion Study Properties* dialog box, enter *1000* for *Frames per second*, and click the checkmark on top. Zoom in the timeline area until you can see tenth second marks. Drag the end time key to the *0.5* second mark (see Figure 5-10) in the timeline area to define the simulation duration.

Click *Calculate* button from the *Motion* toolbar to simulate the motion. A 0.5 second simulation will be carried out.

A warning message appears stating that

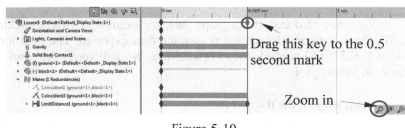

Drag this key to the 0.5 second mark

Zoom in

Figure 5-10

The playback speed or frames per second settings for the motion study will result in poor performance given the current motion study duration. Would you like these settings adjusted for better playback?

Just ignore it (by selecting *No*) since *1000* for *Frames per second* for a 0.5-second will be adequate for our analysis. We may adjust the playback speed if needed after completing the simulation.

Momentarily you should see the block start sliding down on the slope face. You may want to adjust the playback speed to 10% to slow down the animation (by simply choosing *0.1x* from the scroll-down menu next to the animation slider) since the total simulation duration is only 0.5 seconds.

Next, we will graph the linear velocity of the block on the slope face (that is, the *X*-component of the velocity). Thereafter, we graph the friction force between the faces in contact. We will verify the results of *Motion* analysis in Section 5-4.

Plotting Motion Results

We create two graphs, X-velocity of the block when it reaches the end of the slope face, and friction force between the two faces in contact.

In the *MotionManager* tree, right click *block<1>* and select *Create Motion Plot*. In the *Results* dialog box (refer to Figure 5-11a), choose *Displacement/Velocity/Acceleration*, select *Linear Velocity*, and then *X Component*. Note that if you accept the definition for the graph, the results to be graphed will be the sliding velocity of the block along the X-direction of the reference triad (i.e., the global coordinate system) instead of the X-direction of the block that is in parallel with the slope face. This will be our desirable selection.

In the calculation to be carried out externally to *Motion* (see Section 5.4), we calculate the sliding velocity of the block along the slope face, which coincides with the X-direction of the block. Therefore, instead of graphing velocity along the X-direction of the reference triad, we want to graph the velocity along the X-direction of the block. This can be done by clicking the selection field of the *Results* dialog box (circled in Figure 5-11a), and click the origin of the block (make sure you turn on the origin by choosing from the pull-down menu *View → Hide/Show → Origins*). The local coordinate of the block appears in the graphics area with X-axis coincides with the slope face (see Figure 5-11b). Also, the selected object, *block-1*, is listed in the field. Click the checkmark to accept the graph.

A graph like that of Figure 5-12a should appear. From the graph, the *block* takes about 0.34 seconds to reach the end of the slope face. At that point, the X-velocity is 53.559 in/sec in the negative X-direction of the local coordinate system of the block, i.e., moving down the slope face. Note that you may see a slightly different value, which is fine. However, if you see a significantly different value, you may want to make sure *Coincident1* mate is suppressed.

Note that you may double click the vertical axis of the graph to bring up the *Format Axis* dialog box (see Figure 5-12b). From which, you may click the *Number* tab, and enter *3* for *Decimal Places* to increase the decimal points of the X-velocity shown in the graph.

You may also export the graph data; for example, by right clicking the graph and choosing *Export CSV*. Open the spreadsheet and review the data exported. The time for the block to reach the end of the slope is 0.335563 seconds (see Figure 5-13). We will verify the results, including X-velocity and sliding time, in Section 5.4.

Next, we plot the friction force.

From the *Motion* toolbar, click the *Results and Graphs* button ![icon]. In the *Results* dialog box (see Figure 5-14a), choose *Forces*, select *Friction Force*, and then *X Component*. Next, we select the bottom face of the block and then the slope face, as shown in Figure 5-14b).

(a)   (b)

Figure 5-11 Creating Result Graph for the Sliding Velocity of the Block

(a)   (b)

Figure 5-12 The X-velocity Graph of Friction μ_k. = 0.1

Figure 5-13

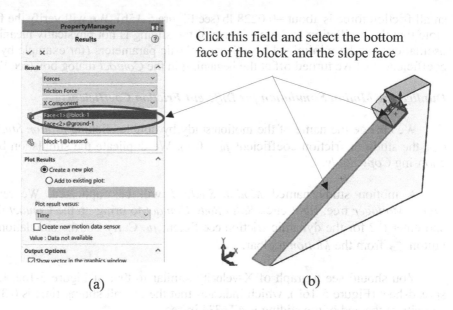

Click this field and select the bottom face of the block and the slope face

(a)   (b)

Figure 5-14 Creating Result Graph for the Friction Force

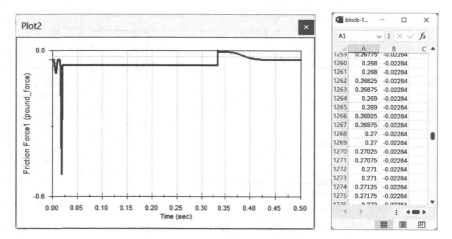

Figure 5-15 The Friction Force Graph and Results of Friction $\mu_k = 0.1$

Similar to that of the X-velocity graph, we click the selection field of the *Results* dialog box (like that circled in Figure 5-11a), and click the block in the graphics area. The local coordinate of the block appears in the graphics area with X-axis in parallel with the slope face (like that of Figure 5-11b). Also, the selected object, *block-1@Lesson5*, is listed in the field (see Figure 5-14a). Click the checkmark to accept the graph. Click *No* for the message:

This motion study has redundant constraints which can lead to invalid force results. Would you like to replace redundant constraints with bushings to ensure valid results? Note that this will make the motion study slower to calculate.

A graph like that of Figure 5-15a should appear. From the graph, except for the oscillation at the beginning, a uniform small friction force is present until the block reaches the end of the slope face at about 0.34 seconds. We export the graph data, open the spreadsheet and review the data. The uniform small friction force is about −0.0228 lb (see Figure 5-15b). We will verify the friction force in Section 5.4. Note that the oscillation at the beginning of the sliding is not physically meaningful and is ignored. Such oscillations can be minimized by adjusting elastic parameters (for example by entering a larger damping coefficient) that we turned off at the beginning in the *Contact* dialog box (see Figure 5-9).

Duplicating Motion Simulation for Different Friction Coefficients

We change the name of the motion study by double clicking *Motion Study 1* tab and entering *mu01* (for the study of friction coefficient $\mu_k = 0.1$). We duplicate the simulation by right clicking *mu01* and choosing *Copy Study*.

A motion study named *Motion Study 1* will be duplicated. We rename it to *mu02*. In the *MotionManager* tree, right click *Solid Body Contact* to bring up the *Contact* dialog box (see Figure 5-9) and enter *0.2* for the dynamic friction coefficient μ_k. Carry out the simulation by clicking the *Calculate* button 🖥️ from the *Motion* toolbar.

You should see a graph of X-velocity similar to that of Figure 5-16a. Exporting the result data in spreadsheet (Figure 5-16b), which indicates that the overall sliding time is 0.377495 seconds, and the X-velocity at the end of the sliding is 47.6334 in/sec.

(a)   (b)

Figure 5-16 The X-velocity Graph of Friction $\mu_k. = 0.2$

Repeat the same steps to create simulations for $\mu_k = 0.3, 0.4, 0.5, 0.01, 0.02, 0.03, 0.05, 0.075$, and 0 (no friction). Note that for the cases of $\mu_k = 0.4$ and 0.5, we need to increase the simulation time to 0.6 and 0.9 seconds, respectively. Record the results of both sliding time and X-velocity at the end of sliding. The results are listed in Table 5-1 under Contact Friction (Obtained from *Motion*). Save your model under a different name.

We have completed the first part of the lesson. We will verify the results of *Motion* analysis in Section 5.4 by formulating equation of motion and solving for the sliding time, velocity at the end, as well as the friction force. Thereafter, we continue the second part of the lesson by defining joint friction as part of the assembly mate. You may want to save your model under a different name and keep the assembly file *Lesson5.SLDASM* unchanged.

5.4 Result Verifications

In this section, we verify analysis results of the simulation obtained from *Motion*. We assume that the *block* is of a concentrated mass so that the particle dynamics theory is applicable.

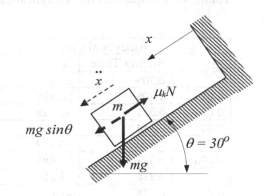

Equation of Motion: Scenario 1

From the free-body diagram shown in Figure 5-17, applying Newton's Second Law for force equilibrium along the x-direction (i.e., along the 30° slope face), we have

Figure 5-17 The Free-Body Diagram

$$\sum F_x = mg\ sin\theta - \mu_k N = m\ddot{x} \qquad (5.1)$$

where $N = mg\ cos\theta$ and μ_k is the kinetic friction coefficient. Therefore,

$$\ddot{x} = g\ sin\theta - \mu_k\ g\ cos\theta \qquad (5.2)$$

where m is the mass of the block, x is the distance between the mass center of the block and the top right end of the slope face, measured from the top right end.

The velocity is obtained by integrating Eq. 5.2 with time and plugging in initial velocity $\dot{x}(0) = 0$ as

$$\dot{x}(t) = g(sin\theta - \mu_k\ cos\theta)\ t \qquad (5.3)$$

One more integration gives the block traveling distance as

$$x(t) = \frac{1}{2}g(sin\theta - \mu_k\ cos\theta)\ t^2 \qquad (5.4)$$

From Eq. 5.4, the total sliding time of the block is

$$t = \sqrt{\frac{2x}{g\ (sin\theta - \mu_k\ cos\theta)}} \qquad (5.5)$$

The X-velocity of the block at the end of the sliding can be obtained by plugging the sliding time obtained from Eq. (5.5) into Eq. (5.3).

Also, the friction force can be obtained by

$$f = \mu_k\ N = \mu_k\ mg\ cos\theta \qquad (5.6)$$

The mass of the block is 0.2637 lbm, which can be obtained by selecting from the pull-down menu *Tools → Evaluate → Mass Properties*, and selecting the block. The analytical calculations can be carried out in a spreadsheet, as listed under *Analytical Calculations* of Table 5-1. As shown in Table 5-1, results obtained from *Motion* analysis, including the sliding time, X-velocity, and friction force, are accurate compared with those of analytical calculations.

Results listed in Table 5-1 can also be found in *Lesson 5 Friction.xls*.

Table 5-1 Comparison for Analytical Calculations and Results Obtained from *Motion* Analysis of Contact Friction

	Analytical Calculations			Contact Friction (Obtained from *Motion*)		
μ_k	Sliding Time	X-velocity	Friction	Sliding Time	X-velocity	Friction
0	0.305	58.948	0	0.305	58.88	0
0.01	0.308	58.435	0.002284	0.308	58.38	0.002284
0.02	0.311	57.918	0.004567	0.311	57.89	0.004568
0.03	0.314	57.396	0.006851	0.313	57.33	0.006852
0.05	0.320	56.337	0.01142	0.319	56.29	0.01142
0.075	0.327	54.986	0.01713	0.327	54.93	0.01713
0.1	0.336	53.600	0.02284	0.336	53.562	0.02284
0.2	0.378	47.656	0.04567	0.378	47.64	0.04568
0.3	0.441	40.856	0.06851	0.440	40.83	0.06852
0.4	0.551	32.671	0.09135	0.551	32.67	0.09136
0.5	0.834	21.576	0.1142	0.834	21.57	0.1142

5.5 Using *SOLIDWORKS Motion*: Joint Friction

We revisit the same block sliding example by specifying friction as part of the joint (or assembly mate).

A joint, or kinematic joint, connects bodies for motion analysis. A joint, depending on its type, allows specific relative motion between the bodies it connects. For example, a pin joint (or hinge joint, revolute joint) allows only rotation motion, and a translational joint (or slider joint) allows only translational motion. The motion allowed, either rotational or translational, is called free degree of freedom (DOF). In motion analysis, a joint produces equal and opposite reactions (forces and/or torques) on the bodies connected at all DOFs, except for free DOF.

Some of the joints are directly equivalent to assembly mate or a combination of assembly mates. For example, mating two planar faces is equivalent to a planar joint. Combining concentric and coincident mates lead to a pin joint (or revolute joint, hinge joint). One thing worth noting is that when mates are combined to create desired free DOF between bodies, they often introduce excessive DOFs known as redundancies. A joint becomes excessive when it does not introduce any further restriction on a body's motion. It is desirable to eliminate redundant DOF whenever possible. The assembly mates defined in *SOLIDWORKS* assembly are converted into kinematic joints in *Motion*. More about kinematic joints, DOF, and redundancies can be found in Appendix A.

As shown in Figure 5-2, in the assembly model, there are three assembly mates, *Coincident1(ground<1>,block<1>)*, *Coincident3(ground<1>,block<1>)*, and *LimitDistance1(ground<1>,block<1>)*. As a result, the block is allowed to move along the slope face.

In defining a motion model, we suppress assembly mate *Coincident3* to eliminate redundant DOF, and we suppress *Coinciden1* and create local coincident mate in *Motion* between the bottom face of the block and the slope face of the ground. Creating local mates will allow changing friction coefficients in *MotionManager* window for the respective motion studies. This local coincident mate is converted into a planar joint in *Motion*.

One important note is that the order of the selected bodies in defining the local coincident mate affects the accuracy of the motion simulation. The right order is block first and then ground, not the other way around. Sometimes *Motion* automatically re-order the selected bodies defined in the mate. The best way to avoid this problem is to re-order the parts in the *FeatureManager* tree of the *SOLIDWORKS* assembly. We will illustrate these important steps next.

We start *SOLIDWORKS* and open assembly file *Lesson5.SLDASM*. We check the units system to make sure that *IPS* is chosen.

Re-ordering ground and the block

In the *FeatureManager* tree, click *block<1>* and drag it up and place it before *ground<1>*. The entity *block<1>* is now placed above *ground<1>*, as shown in Figure 5-18. Again, this is necessary to eliminate the possibility of *Motion* re-ordering the selected bodies included for defining the mate.

Defining Initial Sliding Position

Similar to what we did before, we expand *Mates* in the *FeatureManager* tree, right click *LimitDistance1* and choose *Edit Definition*. We enter *0* in the *LimitDistance1* dialog box (see Figure 5-5) to move the block to the starting position at the top of the slope face.

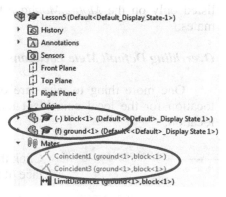

Figure 5-18

Suppressing Coincident Mate: Coincident1 and Coincident3

In the *MotionManager* tree, select *Coincident1* and *Coincident3*, and then right click *Suppress*. The mates should be suppressed and turned into gray color, as shown in Figure 5-18.

Entering Motion

Click the *Motion Study* tab (*Motion Study 1*) at the bottom of the graphics area to bring up the *MotionManager* window.

Defining Gravity

We define gravity pointing in the (negative) *Y* direction like before.

Defining Motion Study-Specific Mate (or Local Mate)

We will define a coincident mate and add friction as a *Motion* study-specific mate (or local mate).

Motion study-specific mates are mates you create for a motion study that are independent of the assembly model mates. Using motion study-specific mates, you can create multiple motion studies to analyze model motion with different mates without changing the assembly solid model. The mates you create when the *MotionManager* window is present become *Motion* study-specific mate automatically.

Click the *Motion Study* tab (*Motion Study 1*) if you have not done so. Then, choose from the pull-down menu *Insert → Mate*. In the *Mate* dialog box (see Figure 5-19a), the *Mate Selections* field will be active, and is ready for you to pick entities.

Pick the bottom face of the block, and then the slope face of the ground, as shown in Figure 5-19a and b. Note again that the order must be correct; that is, block first, and ground second.

The friction coefficient can be entered through the ⚙ *Analysis* tab of the *Mates* dialog box. In the *Mates* dialog box (Figure 5-19c), choose the ⚙ *Analysis* tab, choose *Friction*, enter *0.1* for *Coefficient (μ)*, choose *Planar* for joint, and enter *Length: 1, Width: 1*, and *Radius: 0.7071*.

Note that for calculating friction effects, *Motion* converts the coincident mate between two faces as a planar joint, where the block slides and rotates on the slope face, as illustrated in Figure 5-20, where *L* is the length of the top (sliding) rectangular block, *W* is the width of the top rectangular block, and *R* is the radius of a circle, centered at the center of the block face in contact with the bottom block, which circumscribes the face of the sliding block. For this example, we will enter *L* and *W* to *1*, and *R* to *0.7071*, i.e., $\sqrt{2}/2$.

Click the checkmark ✔ on top of the dialog box to accept the mate definition. A local mate is now added to *MotionManager* tree listed under *Mates* (see Figure 11-21). Motion study-specific mates are listed only on the *MotionManager* tree. The string *Local* prefaces the name of motion study-specific mates.

Overriding Default Mate Location

One more thing to take care of before running a motion analysis is overriding the default mate location for the local coincident mate. We pick the origin of the block to override the default mate location.

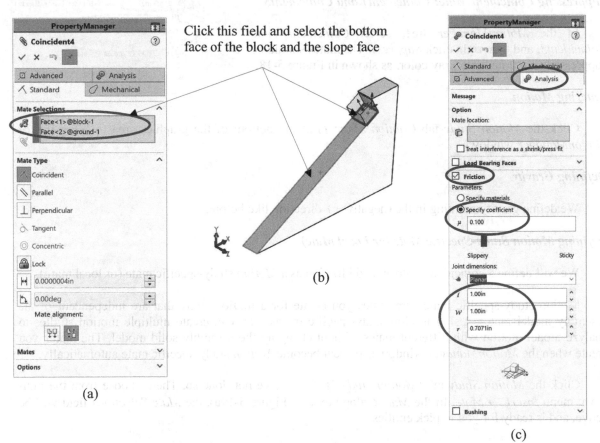

Figure 5-19 Defining a Local Coincident Mate with Friction

From the *MotionManager* tree, right click *Local Coincident* and choose *Edit Feature*. In the *Mates* dialog box, click the *Mate location* field (circled in Figure 5-22a) to activate the location selection, expand the *block<1>* node in the *FeatureManager* tree, and select *Origin* (circled in Figure 5-22b). In the graphics area, the origin of the block selected will be highlighted with a local coordinate system, as shown in Figure 5-22c. Note that if the mate location is not defined properly, the simulation results would not be correct.

Figure 5-20   Figure 5-21

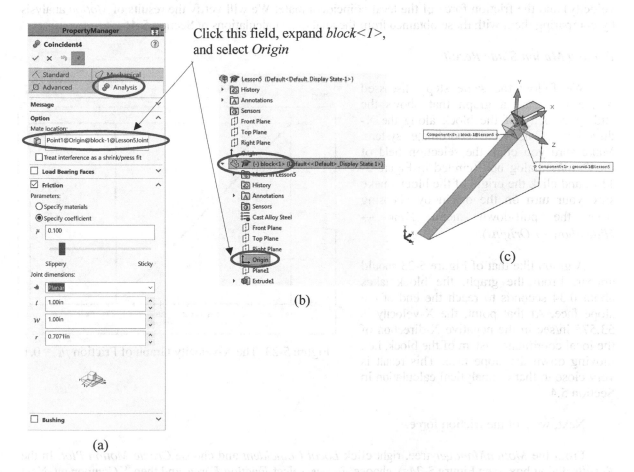

Click this field, expand *block<1>*, and select *Origin*

(a)

(b)

(c)

Figure 5-22   Overriding the Default Mate Location for the Local Coincident Mate

Note that selecting the origin of the block overrides the default mate location. The mate location point determines how parts move in relation to each other. The mate location is always a point. If you select another type of entity such as a face for the mate location, the mate location is at the center of the selected entity.

Defining and Running Simulation

Choose *Motion Analysis* for study. Click the *Motion Study Properties* button ⚙ from the *Motion* toolbar. In the *Motion Study Properties* dialog box, enter *1000* for *Frames per second*, and click the checkmark on top. Zoom in the timeline area until you can see the tenth second marks. Drag the end time key to *0.5* second mark (like that of Figure 5-10) in the timeline area to define the simulation duration.

Click *Calculate* button 🖩 from the *Motion* toolbar to simulate the motion. A 0.5 second simulation will be carried out.

Again, if you see a warning message stating: *The playback speed or frames per second settings for the motion study will result in poor performance given the current motion study duration*, just ignore it (by selecting *No*) for the same reasons stated earlier.

Next, we will graph the linear velocity of the block on the slope face (that is, *X*-component of the velocity) and the friction force of the local coincident mate. We will verify the results of *Motion* analysis by comparing them with those obtained from the analytical calculations of Section 5-4.

Plotting Motion Study Results

We follow the same steps discussed before to create a graph that shows the sliding velocity of the block along the X-direction of its local coordinate system. Make sure you click the selection field of the *Results* dialog box (circled in Figure 5-11a), and click the origin of the block (make sure your turn on the origin by choosing from the pull-down menu *View* → *Hide/Show* → *Origins*).

A graph like that of Figure 5-23 should appear. From the graph, the block takes about 0.34 seconds to reach the end of the slope face. At that point, the X-velocity is 53.575 in/sec in the negative X-direction of the local coordinate system of the block, i.e., moving down the slope face. This result is very close to that of analytical calculation in Section 5.4.

Figure 5-23 The X-velocity Graph of Friction $\mu_k = 0.1$

Next, we plot the friction force.

From the *MotionManager* tree, right click *Local Coincident* and choose *Create Motion Plot*. In the *Results* dialog box, (see Figure 5-24a), choose *Forces*, select *Friction Force*, and then *X Component*. Next, we select the origin of the block as the reference part. Similar to that of the X-velocity graph, we click the selection field of the *Results* dialog box (circled in Figure 5-24a), and click the origin of the block in the graphics area. The local coordinate of the block appears in the graphics area with X-axis in parallel with the slope face (like that of Figure 5-24b). Also, the selected object, *block-1@Lesson5*, is listed in the field. Click the checkmark to accept the graph.

Click *Yes* if you see a warning message: *This motion study has redundant constraints which can lead to invalid force results. Would you like to replace redundant constraints with bushings to ensure valid force results? Note that this will make the motion study slower to calculate.*

A graph like that of Figure 5-24c should appear. From the graph, except for the jump at the beginning, a uniform small friction force is present until the block reaches the end of the slope face at about 0.34 seconds.

We export the graph data, open the spreadsheet and review the data exported. The uniform small friction force is about 0.02282 lb (see Figure 5-24d), which is close to that of analytical calculations discussed in Section 5.4.

Figure 5-24 Creating Result Graph for the Friction Force

Duplicating Motion Simulation for Different Friction Coefficients

Similar to the first part of the lesson, we change the name of the simulation to *mu01* and then duplicate the motion study for different friction coefficients.

Note that you will have to repeat the steps illustrated in Figure 5-19 to enter the friction coefficient and joint dimensions. Moreover, we will make sure the block origin is selected as the mate location, as illustrated in Figure 5-22.

We duplicate the motion study for μ_k = 0.2, 0.3, 0.4, 0.5, 0.01, 0.02, 0.03, 0.05, 0.075, and 0 (no friction). Note again that for the cases of μ_k = 0.4 and 0.5, we need to increase the simulation time to 0.6 and 0.9 second, respectively. Record the results of both sliding time and X-velocity at the end of sliding.

The results are listed in Table 5-2 under Joint Friction (Obtained from *Motion*). As shown in Table 5-2, results obtained from *Motion* analysis, including the sliding time, X-velocity, and friction force, are accurate comparing with those of analytical calculations.

Note that you may see a friction force graph showing negative friction forces, like that of Figure 5-25a (in this case, the friction coefficient is $\mu = 0.2$). This is because for some reason *Motion* changes the order of the two faces selected for defining the coincident mate. In Figure 5-25b, Face<1>@block<1> is listed before Face<2>@ground<1>. In this case, the friction forces are graphed as positive value like that of Figure 5-24c. In Figure 5-25c, Face<1>@ground<1> is listed before Face<2>@block<1>, which leads to a graph showing negative friction forces, like that of Figure 5-25a.

Table 5-2 Comparison for Analytical Calculations and Results Obtained from *Motion* Analysis of Joint Friction

μ_k	Analytical Calculations			Joint Friction (Obtained from *Motion*)		
	Sliding Time	X-velocity	Friction	Sliding Time	X-velocity	Friction
0	0.305	58.948	0	0.305	58.88	0
0.01	0.308	58.435	0.002284	0.307	58.24	0.002278
0.02	0.311	57.918	0.004567	0.310	57.78	0.004565
0.03	0.314	57.396	0.006851	0.313	57.30	0.006846
0.05	0.320	56.337	0.01142	0.319	56.27	0.01141
0.075	0.327	54.986	0.01713	0.327	54.96	0.01713
0.1	0.336	53.600	0.02284	0.335	53.58	0.02282
0.2	0.378	47.656	0.04567	0.376	47.60	0.04554
0.3	0.441	40.856	0.06851	0.438	40.88	0.06838
0.4	0.551	32.671	0.09135	0.544	32.74	0.09121
0.5	0.834	21.576	0.1142	0.801	21.81	0.1140

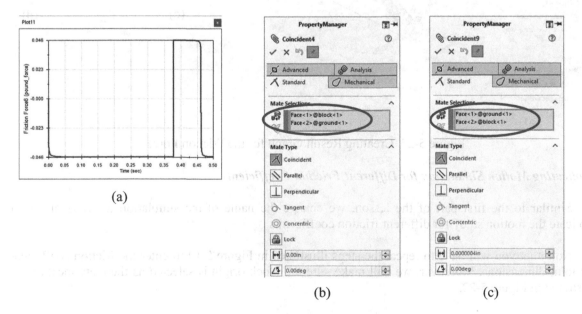

(a)

(b)

(c)

Figure 5-25 Negative Friction Forces

We have completed the lesson. Save your model under a different name for future reference.

Exercises:

1. Use the two parts provided in the Exercise 5-1 folder, block and base, to create an assembly like that of Figure E5-1. Add a downward force of 100N on the top face of the block, and a linear motor for the block at a constant speed of 200 mm/s along the positive X-direction. Define a local coincident mate between the bottom face of the block and the top face of the base. Specify a friction coefficient of 0.25. Run a *Motion* study for 2 seconds. Plot *Motion* study results for the friction force, reaction force, and motor force. Calculate the friction force externally and compare the force calculated with that obtained from *Motion* study.

Figure E5-1

2. Repeat Problem 1 by defining a body contact between the block and the base (instead of a local coincident mate).

Lesson 6: A Spring-Mass System

6.1 Overview of the Lesson

In this lesson, we create a simple spring-mass system using an example similar to *Lesson 5* and simulate its dynamic responses under two scenarios. A schematic of the system is shown in Figure 6-1, in which a steel block of *1*in.×*1*in.×*1*in. is sliding along a *30°* slope face with a spring connecting it to the top end of the slope face. The block will slide back and forth on the slope face under two scenarios. First, the block will slide due to a small initial displacement, essentially, a free vibration problem. For the second scenario, we add a sinusoidal force *p(t)*; therefore, it is a forced vibration problem. Gravity will be turned on for both scenarios. In this lesson, we will learn how to create the spring-mass model, run a motion analysis, and visualize the analysis results. The analysis results of the spring-mass example can be verified using particle dynamics theory. Similar to previous lessons we formulate the equation of motion, solve the equations, graph positions of the block, and compare our calculations with results obtained from *Motion*.

6.2 The Spring-Mass System

Physical Model

Note that the *IPS* units system will be used for this example. The spring constant and the unstretched length (or free length) are $k = 20$ lb$_f$/in. and $U = 3$ in., respectively. As mentioned earlier, the first scenario assumes a free vibration, where the block is stretched *1* in. downward along the *30°* slope face. In the second scenario, an external force $p(t) = 10 \cos 2\pi t$ lb$_f$ is applied to the block, as shown in Figure 6-1. Both scenarios assume a gravity of $g = 386$ in/sec^2 in the negative *Y*-direction. Both will be simulated using *Motion*.

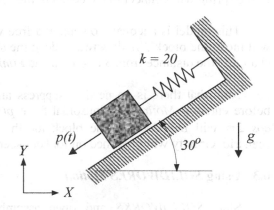

Figure 6-1 The Spring-Mass System

SOLIDWORKS Part and Assembly Files

For this lesson, the parts and assembly have been created in *SOLIDWORKS*. There are four model files created: *block.SLDPRT*, *ground.SLDPRT*, *Lesson6.SLDASM*, and *Lesson6withResults.SLDASM*. We will start with *Lesson6.SLDASM*, in which the block is assembled to the ground and no motion entities have been added. The assembly file *Lesson6withResults.SLDASM* contains the complete simulation models with simulation results for both scenarios. You may open this file and review motion simulation results before going over this lesson.

In the assembly models, there are three assembly mates identical to those of *Lesson 5*, *Coincident1(ground<1>,block<1>)*, *Coincident3(ground<1>,block<1>)*, and *LimitDistance1(ground<1>,block<1>)*, as shown in Figure 6-2. The block is allowed to move along the slope face.

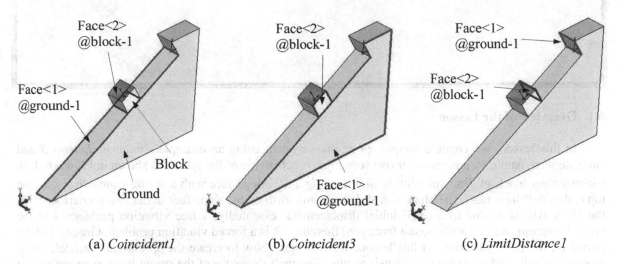

(a) *Coincident1*   (b) *Coincident3*   (c) *LimitDistance1*

Figure 6-2 Assembly Mates Defined in *Lesson6.SLDASM*

Motion Model

A spring with a spring constant $k = 20$ lb$_f$/in. and an unstretched length $U = 3$ in. will be added to connect the block (*Face<2>*, as shown in Figure 6-2c) with the ground (*Face<1>*). By default, the ends of the spring will connect to the center of the corresponding square faces (see Figure 6-3).

This model is adequate to support a free vibration simulation under the first scenario. Note that we will move the block *1*in. downward along the slope face for the simulation. This can be accomplished by changing the distance from *3* to *4* in. in the *LimitDistance1* assembly mate.

Note that there is no need to suppress any mate before entering *Motion*. A sinusoidal force $p(t) = 10 \cos 2\pi t$ will be added to the block for the second scenario. Gravity will be turned on for both scenarios.

6.3 Using *SOLIDWORKS Motion*

Start *SOLIDWORKS* and open assembly file *Lesson6.SLDASM*.

Before creating any motion entities, always check the units system. Similar to *Lesson 3*, choose from the pull-down menu

Tools → *Options*

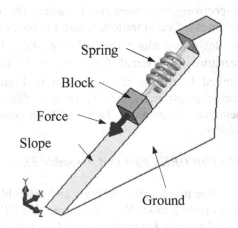

Figure 6-3 Spring Mass Dynamic Model

and choose the *Document Properties* tab in the *Document Properties-Units* dialog box, then click the *Units* node. The *IPS* should have been selected. Close the dialog box. In this units system, the gravity is *386* in/sec² in the negative *Y*-direction of the global coordinate system.

Click the *Motion Study* tab (*Motion Study 1*) at the bottom of the graphics area to bring up the *MotionManager* window.

Defining Gravity

Click the *Gravity* button ⌀ from the *Motion* toolbar to bring up the *Gravity* dialog box. Choose *Y* and keep the g-value (386.09in/s^2) in the dialog box, as shown in Figure 6-4. In the graphics area, an arrow appears at the right lower corner ⬇, pointing downward indicating the direction of the gravity.

Click the checkmark ✔ on top of the dialog box to accept the gravity. A gravity node (name: *Gravity*) should appear in the *MotionManager* tree.

Figure 6-4

Defining Spring

Click the *Spring* button ☰ from the *Motion* toolbar to bring up the *Spring* dialog box. In the *Spring* dialog box (Figure 6-5), choose *Linear Spring* (default). The empty field right underneath the *Spring parameters* label is active for you to pick entities to define the two ends of the spring. Pick the face at top right of the ground (see Figure 6-6). Rotate the view, and then pick the face in the block, as shown in Figure 6-6. A spring symbol should appear in the graphics area, connecting the center points of the two selected faces.

Figure 6-5 The *Spring* Dialog Box

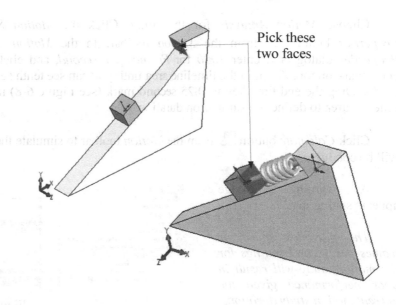

Pick these two faces

Figure 6-6 Picking Two Faces to Define the Spring

In the *Spring* dialog box (Figure 6-5), enter the followings:

Stiffness: 20
Length: 3 (Note that you have to deselect the *Update to model changes* box before entering this value)
Coil Diameter: 0.75
Number of Coils: 5
Wire Diameter: 0.25

Click the checkmark to accept the definition and close the *Spring* dialog box. A spring node (name: *LinearSpring1*) should appear in the *MotionManager* tree.

Defining Initial Position

We would like to stretch the spring *1*in. downward along the slope face as the initial position for the block. The block will be released from this position to simulate a free vibration, i.e., the first scenario. We will edit the mate *LimitDistance1*; enter *4* for the distance dimension.

From the *SOLIDWORKS* browser, expand the *Mates* node listed in the browser, right click *LimitDistance1(ground<1>,ball<1>)*, and choose *Edit Feature*. You should see the definition of the assembly mate in the dialog box like that of Figure 6-7. Enter *4.00* in. for the distance, and click the checkmark on top to accept the change. In the graphics area, the block should move to the *4.00* in. position on the slope face.

Defining and Running Simulation

Choose *Motion Analysis* for the study. Click the *Motion Study Properties* button from the *Motion* toolbar. In the *Motion Study Properties* dialog box, enter *1000* for *Frames per second*, and click the checkmark on top. Zoom in the timeline area until you can see tenth second marks. Drag the end time key to 0.25 second mark (see Figure 6-8) in the timeline area to define the simulation duration.

Figure 6-7

Click *Calculate* button from the *Motion* toolbar to simulate the motion. A 0.25 second simulation will be conducted.

A warning message appears stating that

The playback speed or frames per second settings for the motion study will result in poor performance given the current motion study duration.

Drag this key to the 0.25 second mark

Zoom in

Figure 6-8

Just ignore it by selecting *No* since *1000* for *Frames per second* for a 0.25-second will be adequate for our analysis. We may adjust the playback speed if needed after completing the simulation.

After a brief moment, you should see the block start moving back and forth on the slope face. You may want to adjust the playback speed to 10% to slow down the animation (by simply choosing *0.1x* from the scroll-down menu next to the animation slider) since the total simulation duration is only 0.25 seconds.

Next, we will graph the spring length which describes the displacement of the block on the slope face (instead of *X*- or *Y*-component). This displacement graph should reveal a sinusoidal function as we have seen in many vibration examples of *Physics*.

Displaying Simulation Results

Click the *Results and Plots* button from the *Motion* toolbar. In the *Results* dialog box (refer to Figure 6-9), choose *Displacement/Velocity/Acceleration*, select *Linear Displacement*, and then *Magnitude*. Similar to the spring, pick the face at top right of the ground (see Figure 6-6). Rotate the view, and then pick the face in the block, as shown in Figure 6-6. Click the checkmark to accept the graph.

A graph like that of Figure 6-10 should appear. From the graph, the *block* moves on top of the slope face between *2* and *4* in. This is because the unstretched length of the spring is *3* in. and we stretched the spring *1* in. to start the motion. Also, it takes about *0.04* seconds to complete a cycle, which is small. The small vibration period can be attributed to the fact that the spring is fairly stiff (*20* lb$_f$/in).

Note that if you see a graph similar to Figure 6-10 but with upper and lower bounds 4.5 and 2.5 in., respectively, the graph would be incorrect. The software bug was reported to SOLIDWORKS company on 1/18/2010 (SR 1-1493146001). You may consider selecting two vertices (see Figure 6-11), instead of faces, for defining the displacement graph.

Note that you can also export the graph data, for example, by right clicking the graph and choosing *Export CSV*. Open the spreadsheet and review the data exported. The time for the block to move back to its initial position, i.e., when the distance is *4* in., is *0.037* seconds.

We will conduct analytical calculations to verify these results later in Section 6.4. Before we do that, we will work on the second scenario: with the addition of an external force.

Double click the *Motion Study 1* tab, and change the name to *Scenario 1*. Save the model before moving to the next scenario.

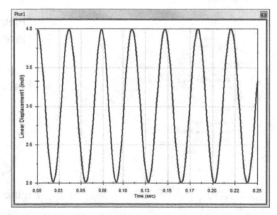

Figure 6-9 The *Results* Dialog Box   Figure 6-10 The Displacement Graph: *Scenario 1*

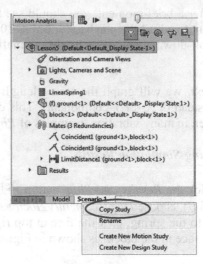

Pick these two vertices

Figure 6-11 Pick Vertices for Plot

Figure 6-12

Scenario 2: With External Force p(t)

In this scenario we will add an external force $p(t) = 10 \cos 2\pi t$ at the center of the end face of the *block* in the downward direction along the slope face.

Copy the motion study and change its name to *Scenario 2* (see Figure 6-12).

The force can be added by choosing the *Force* button ↖ from the *Motion* toolbar. In the *Force* dialog box (Figure 6-13), choose *Linear Force* for *Force Type* and *Action Only* (default) for *Direction*, and then pick the left end face of the block, as shown in Figure 6-14, for the location of the applied force.

The picked face *Face<1>@block-1* is now listed. An arrow appears in the downward direction (shown in Figure 6-14), indicating the direction of the force. That is, the force will be applied to the center of the end face and in the downward direction that is normal to the selected face.

Figure 6-13 The *Force* Dialog Box

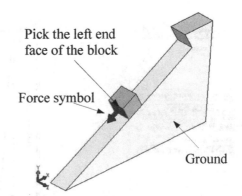

Pick the left end face of the block

Force symbol

Ground

Figure 6-14 Pick Face to Defining Location and Direction of the Force

Scroll down the dialog box to define *Force Function*. Click the *Function* tab (see Figure 6-13) and choose *Expression*. A *Function Builder* dialog box appears (see Figure 6-15). Enter *10*COS(2*PI*time)*, deselect *Force derivative*, enter *0* and *0.5* for *Minimum x value* and *Maximum x value*, respectively (all circled in Figure 6-15). You should see a cosine function graphed like that of Figure 6-15. Click *OK* to accept the force function.

Figure 6-15 The *Function Builder* Dialog Box

Click checkmark ✔ on top of the *Force/Torque* dialog box to accept the force definition. A force node (name: *Force1*) should appear in the *MotionManager* tree.

Next, we change the simulation duration to *0.5* seconds by dragging the end time key to the 0.5-second mark. Note that the *0.5*-second duration is half the harmonic function period of the force applied to the block.

Click the *Calculate* button 🖩 to run a simulation (with the same simulation parameters as those of *Scenario 1*, except with duration extended to 0.5 seconds). After a brief moment, the block starts moving.

As soon as the simulation is completed, a graph like that of Figure 6-16 should appear. From the graph, the *block* moves up and down the slope face roughly for *2*in. The vibration amplitude is enveloped by a cosine function due to the external force *p(t)*. Also, it takes about *0.04* seconds to complete a cycle, which is unchanged from the previous scenario. We will conduct our own calculations to verify these results later. Save the model.

Note that if you see a graph that shows a sinusoidal curve similar to that of Figure 6-16, but with increasing amplitude, the result is incorrect. One possible remedy is to increase the accuracy to 0.00001 (default is 0.0001) from the *Motion Study Properties* dialog box, as shown in Figure 6-17, by clicking the *Motion Study Properties* button ⚙ from the *Motion* toolbar.

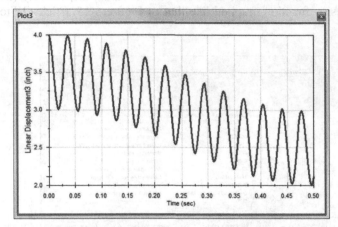

Figure 6-16 The Displacement Graph: *Scenario 2*

Figure 6-17

6.4   Result Verifications

In this section, we will verify analysis results of both scenarios obtained from *Motion*. We will assume that the *block* is of a concentrated mass so that the particle dynamics theory is applicable.

We will start with *Scenario 1* (i.e., free vibration with gravity), and then solve the equations of motion for *Scenario 2* (forced vibration).

Equation of Motion: Scenario 1

From the free-body diagram shown in Figure 6-18, applying Newton's Second Law and force equilibrium along the *X*-direction (i.e., along the 30° slope), we have

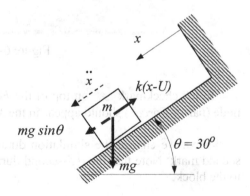

Figure 6-18 The Free-Body Diagram

$$\sum F_x = mg\ sin\theta - k(x-U) = m\ddot{x} \qquad (6.1)$$

Therefore,

$$m\ddot{x} + k(x-U) = mg\ sin\theta \qquad (6.2)$$

where m is the mass of the block, U is the unstretched length of the spring, x is the distance between the mass center of the block and the top right end of the slope face, measured from the top right end.

The double dots on top of x represent the second derivative of x with respect to time. Rearranging Eq. 6.2, we have

$$m\ddot{x} + kx = mg\ sin\theta + Uk \qquad (6.3)$$

where both terms on the right are time independent.

This is a second-order ordinary differential equation. It is well known that the general solution of the differential equation is:

$$x_g = A_1 \cos \omega_n t + A_2 \sin \omega_n t \tag{6.4}$$

where $\omega_n = \sqrt{\dfrac{k}{m}}$, and A_1 and A_2 are constants to be determined with initial conditions. Note that the mass of the steel block is *0.264 pounds* (this is pound mass, lb_m). This can be obtained from *SOLIDWORKS* by selecting the block, and choosing, from the pull-down menu, *Tools → Evaluate → Mass Properties* from the *Mass Properties* dialog box (see Figure 6-19).

Figure 6-19 The *Mass Properties* Dialog Box

Figure 6-20 The Excel Spreadsheet

Note that there are *2* decimal points set in *SOLIDWORKS* by default. You may increase it by clicking the *Options* button, or using the *Document Properties – Units* dialog box (choose from pull-down menu *Tools → Options*).

Note that the pound-mass unit lb_m is not as common as the slug with which we are familiar. The corresponding force unit of lb_m is lb_m in/sec² according to Newton's Second Law.

Also, a *1-lb_m* mass block weighs *1* lb_f on earth; therefore, *1 lb_f = 386 lb_m in/sec²*. The stiffness we employed for the spring becomes k = *20* lb_f/in = *20×386* lb_m in/sec². Hence, $\omega_n = \sqrt{\dfrac{k}{m}} = \sqrt{\dfrac{20 \times 386}{0.264}} = 171.0$ rad/sec = *9,798* degrees/sec, and the natural frequency of the system is f_n = $\omega_n/2\pi$ = *27.2* Hz. The period for a complete vibration cycle is $T = 1/f_n = 0.0367$ seconds, which is close to what was obtained by reviewing the graph (for example, the graph shown in Figure 6-10) and spreadsheet data converted from the graph.

For more details regarding the force and mass in the English units system, please refer to Appendix B. Appendix B should clarify some of the confusion you might have in lb_m and lb_f.

The particular solution of Eq. 6.3 is

$$x_p = \frac{mg \sin\theta}{k} + U \tag{6.5}$$

Therefore, the total solution is

$$x = x_g + x_p = A_1 \cos \omega_n t + A_2 \sin \omega_n t + \frac{mg \sin\theta}{k} + U \tag{6.6}$$

The initial conditions for the spring-mass system are $x(0) = x_0 = 4$ in., and $\dot{x}(0) = 0$ in/sec. Plugging the initial conditions into Eq. 6.6 (you will have to take the derivative of Eq. 6.6 for $\dot{x}(0)$), we have

$$A_1 = x_0 - \left(\frac{mg \sin\theta}{k} + U\right), \text{ and}$$

$$A_2 = 0.$$

Hence, the overall solution is

$$x = \left(x_0 - \frac{mg \sin\theta}{k} - U\right) \cos \omega_n t + \frac{mg \sin\theta}{k} + U \tag{6.7}$$

Figure 6-21 Solution from Theory: *Scenario 1*

Equation 6.7 can be implemented into an Excel spreadsheet, as shown in Figure 6-20. Column B in the spreadsheet shows the results of Eq. 6.7, which is graphed in Figure 6-21. Comparing Figure 6-21 with Figure 6-10, the results obtained from theory and *Motion* agree very well, which means the motion model has been properly defined, and *Motion* gives us reliable results.

Equation of Motion: Scenario 2

Refer to the free-body diagram shown in Figure 6-18 again. For *Scenario 2* we must include the force $p = f_0 \cos(\omega t)$ along the X-direction for force equilibrium, i.e.,

$$m\ddot{x} + k(x - U) = mg \sin\theta + f_0 \cos(\omega t) \tag{6.8}$$

where $f_0 = 10$ lb_f and $\omega = 2\pi$ rad/sec. Rearranging Eq. 6.8, we have

$$m\ddot{x} + kx = mg \sin\theta + Uk + f_0 \cos(\omega t) \tag{6.9}$$

where the right-hand side consists of constant and time-dependent terms.

For the constant terms, the particular solution is identical to that of *Scenario 1*, i.e., Eq. 6.5. For the time-dependent term, $p = f_0 \cos(\omega t)$, the particular solution is

$$x_{p_2} = \frac{f_0}{k - \omega^2 m} \cos \omega t \qquad (6.10)$$

Therefore, the overall solution of Eq. 6.9 becomes

$$x = x_g + x_p + x_{p_2} = A_1 \cos \omega_n t + A_2 \sin \omega_n t + \frac{mg \sin \theta}{k} + U + \frac{f_0}{k - \omega^2 m} \cos \omega t \qquad (6.11)$$

Plugging the initial conditions into the solution, we have

$$A_1 = x_0 - \left(\frac{mg \sin \theta}{k} + U + \frac{f_0}{k - \omega^2 m} \right), \text{ and } A_2 = 0.$$

Hence, the complete solution is

$$\begin{aligned}
x &= \left(x_0 - \frac{mg \sin \theta}{k} - U - \frac{f_0}{k - \omega^2 m} \right) \cos \omega_n t + \frac{mg \sin \theta}{k} + U + \frac{f_0}{k - \omega^2 m} \cos \omega t \\
&= \left[\left(x_0 - \frac{mg \sin \theta}{k} - U \right) \cos \omega_n t + \frac{mg \sin \theta}{k} + U \right] + \frac{f_0}{k - \omega^2 m} (\cos \omega t - \cos \omega_n t)
\end{aligned} \qquad (6.12)$$

Note that terms grouped in the first bracket of Eq. 6.12 are identical to those of Eq. 6.7, i.e., *Scenario 1*. The second term of Eq. 6.12 graphed in Figure 6-22 represents the contribution of the external force *p(t)* to the block motion. The graph shows that the amplitude of the block is kept within *1* in., but the position of the block varies in time. The vibration amplitude is enveloped by a cosine function.

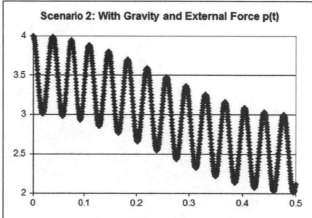

Figure 6-22 Graph of the Second Term of Eq. 6.12

Figure 6-23 Solution from Theory: Scenario 2

The overall solution of *Scenario 2*, i.e., Eq. 6.12, is a combination of graphs shown in Figures 6-21 and 6-22. In fact, Eq. 6.12 has been implemented in Column C of the spreadsheet. The data are graphed in Figure 6-23.

Comparing Figure 6-23 with Figure 6-16, the results obtained from theory and *Motion* are close. Note that the spreadsheet shown in Figure 6-20 can be found at the publisher's website (filename: *lesson6.xls*).

Exercises:

1. Show that Eq. 6.12 is the correct solution for *Scenario 2* governed by Eq. 6.8 by simply plugging Eq. 6.12 into Eq. 6.8.

2. Repeat the *Scenario 2* of this lesson, except changing the external force to $p(t) = 10 \cos 9798.0t$ lb$_f$. Will this external force change the vibration amplitude of the system? Can you simulate this resonance scenario in *Motion*?

3. Add a damper with damping coefficient $C = 0.01$ lb$_f$ sec/in. and repeat the *Scenario 1* simulation using *Motion*.

    (i)   Calculate the natural frequency of the system and compare your calculation with that of *Motion*.

    (ii) Derive and solve the equations that describe the position and velocity of the mass. Compare your solutions with those obtained from *Motion*.

Notes:

Lesson 7: A Slider-Crank Mechanism

7.1 Overview of the Lesson

In this lesson, we learn how to create simulation models for a slider-crank mechanism and conduct three analyses: kinematic, interference, and dynamic. You will learn how to select assembly mates to connect parts in order to create a successful motion model. We will first drive the mechanism by rotating the crank with a constant angular velocity; basically, conducting a kinematic analysis. After we complete a kinematic analysis, we will turn on interference checking and repeat the analysis to see if parts interfere or collide. It is important to make sure no interference exists between parts while the mechanism is in motion. Finally, we will go over a dynamic analysis where we will add a firing force to the piston. This lesson will start with a brief overview about the slider-crank assembly created in *SOLIDWORKS*. At the end of this lesson, we will verify the kinematic simulation results using theory and computational methods commonly found in mechanism design textbooks.

7.2 The Slider-Crank Example

Physical Model

The slider-crank mechanism is essentially a four-bar linkage, as shown in Figure 7-1. They are commonly found in mechanical systems; e.g., internal combustion engine and oil-well drilling equipment. For the internal combustion engine, the mechanism is driven by a firing load that pushes the piston, converting the reciprocal motion into rotational motion at the crank.

Figure 7-1 Schematic View of the Slider-Crank Mechanism

In the oil-well drilling equipment, a torque is applied at the crank. The rotational motion is converted to a reciprocal motion at the piston that digs into the ground. Note that in any case the length of the crank must be smaller than that of the rod in order to allow the mechanism to operate properly. This is commonly referred to as Grashof's law. In this example, the lengths of the crank and rod are 3 in. and 8 in., respectively.

Note that the units system chosen for this example is *IPS* (*in-lb$_f$-sec*). All parts are made of Aluminum, *2014 Alloy*. No friction is assumed between any pair of the components (parts or subassemblies).

SOLIDWORKS Part and Assembly Files

The slider-crank system consists of five parts and one subassembly. They are bearing, crank, rod, pin, piston, and rodandpin (subassembly, consisting of rod and pin). An explode view of the mechanism is shown in Figure 7-2. *SOLIDWORKS* parts and assembly have been created for you. They are *bearing.SLDPRT*, *crank.SLDPRT*, *rod.SLDPRT*, *pin.SLDPRT piston.SLDPRT*, and *rodandpin.SLDASM*. In addition, there are two assembly files: *Lesson7.SLDASM* and *Lesson7withResults.SLDASM*. You can find these files at the publisher's web site.

We will start with *Lesson7.SLDASM*, in which all components are properly assembled. In this assembly the bearing is anchored (ground) and all other parts are fully constrained. We will suppress one assembly mate before entering *Motion* in order to allow for movement.

Same as before, the assembly file *Lesson7withResults.SLDASM* consists of a complete simulation model with simulation results. You may want to open this file to see the motion simulation of the mechanism. In the assembly file with complete simulation results, a mate (*Concentric2*) has been suppressed. You can also see how the parts move by moving the mouse pointer to the graphics area and dragging a movable part, for example the crank. The whole mechanism will move accordingly.

Figure 7-2 Slider-Crank Assembly (Explode View)

There are eight assembly mates, including five coincident and three concentric, defined in the assembly. You may want to expand the *Mates* node listed in the *FeatureManager* tree to review the assembly mates. Click any of the mates; you should see the geometric entities selected for the assembly mate highlighted in the graphics area.

The first three mates (*Concentric1*, *Coincident1*, and *Coincident2*) assemble the crank to the fixed bearing, as shown in Figure 7-3a. As a result, the crank is completely fixed. Note that the mate *Coincident2* orients the crank to the upright position, defining the initial configuration of the mechanism. This mate will be suppressed before entering *Motion*. Suppressing this mate will allow the crank to rotate with respect to the bearing. Note that combining these two mates, *Concentric1* and *Coincident1*, yields a revolute joint (also called pin or hinge joint). However, after *Motion 2008* only assembly mates are employed for motion models. Kinematic joints, such as revolute joint, cylindrical joint, etc. commonly discussed in kinematic and dynamic textbooks, cannot be found in *Motion 2024*. If interested, you may refer to Appendix A for more discussion on the subject of kinematic joints.

The next two mates (*Concentric2* and *Coincident3*) assemble the rod to the crank, as shown in Figure 7-3b. Unlike the crank, the rod is allowed to rotate with respect to the crank. The next two mates (*Concentric3* and *Coincident4*) assemble the piston to the pin, allowing the piston to rotate about the pin. The final mate (*Coincident5*) eliminates the rotation by mating two planes, *Right Plane* of the piston and the *Top Plane* of the assembly, as shown in Figure 7-3c.

(a) Assemble Mates for Crank (Explode View)

(b) Assemble Mates for Rod (Explode View)

(c) Assemble Mates for Piston (Unexplode View)

Figure 7-3 Assembly Mates Defined for the Mechanism

Simulation Model

Once we suppress the mate *Coincident2*, the mechanism has one free degree of freedom. Either rotating the crank or the rod (about the axis of the concentric mates, i.e., parallel to the *Z*-axis), or moving the piston horizontally (along the *X*-direction) will be sufficient to uniquely determine the position, velocity, and acceleration of any parts in the mechanism. Kinematically this mechanism is identical to that of the single piston engine presented in *Lesson 1*. We know physically the total degree of freedom of the system is *1* before adding a rotary motor. However, if we carry out a Gluebler's count, we will have the following:

3 (moving bodies) × *6 (DOFs/body)* − *3 (concentric)* × *4 (DOFs/concentric)* − *4 (coincident)* × *3 (DOFs/coincident)* = *18 −24= −6*

Apparently, there are seven redundant DOFs embedded in the motion model. Since *Motion* will automatically detect and remove redundant DOFs during motion simulation, we will not worry too much about redundancy. *Motion* will report the number of redundancies after a motion analysis is carried out. We will see this later.

The mechanism will be first driven by rotating the crank at a constant angular velocity of *360* degrees/sec, as shown in Figure 7-4. Gravity will be turned off. This will be essentially a kinematic simulation. This model will also serve for interference checking. It is important for designers to make sure no interference exists between parts while the mechanism is in motion. The simulation results are included in *Lesson7withResults.SLDASM* under *Kinematic* motion study tab.

Figure 7-4 The Simulation Model

The next and final analysis will be dynamic, where we will add a firing force to the piston for a dynamic simulation. The results are included in *Lesson7withResults.SLDASM* under *Dynamic* motion study tab.

7.3   Using *SOLIDWORKS Motion*

Start *SOLIDWORKS* and open assembly file *Lesson7.SLDASM*.

First, suppress the third assembly mate *Coincident2*. From the *Assembly* browser, expand the *Mates* node, right click *Coincident2*, and choose *Suppress* ⬇⬚ . The mate *Coincident2* becomes inactive.

Before creating any entities, always check the units system. Make sure *IPS* units system is chosen for this example.

Click the *Motion Study* tab at the bottom of the graphics area to bring up the *MotionManager* window. Double click the tab and rename the study *Kinematic*.

We will add a rotary motor to drive the crank and conduct a 1-second kinematic analysis. Later we will suppress the rotary motor and add a force for a dynamic simulation.

Adding a Rotary Motor

Click the *Motor* button ⟳ from the *Motion* toolbar to bring up the *Motor* dialog box (Figure 7-5). Choose *Rotary Motor* (default). Move the pointer to the graphics area, and pick a circular arc of the crank to define the rotation direction of the rotary motor; e.g., the circular edge on the lower shaft of the crank, as shown in Figure 7-6. A circular arrow appears indicating the rotational direction of the rotary motor. A counterclockwise direction (Z-direction) is desired. You may change the direction by clicking the direction button ⟲ below *Component/Direction*. Choose *Constant speed* and enter *60* RPM for speed. Click the checkmark ✔ on top of the dialog box to accept the motor definition. You should see a rotary motor *RotaryMotor1* added to the *MotionManager* tree.

Figure 7-5

Pick this Arc

Figure 7-6

Figure 7-7 Motion Simulation

Figure 7-8

Defining and Running Simulation

Choose *Motion Analysis* for the study. Click the *Motion Study Properties* button ⚙ from the *Motion* toolbar. In the *Motion Study Properties* dialog box, enter *200* for *Frames per second*, and click the checkmark on top. Drag the end time key to one-second mark in the timeline area to define the simulation duration. Zoom in to the timeline area if needed.

Click the *Calculate* button ⊞ from the *Motion* toolbar to simulate the motion. A one-second simulation will be carried out. After a short moment, you should see that the mechanism starts moving. The crank rotates *360* degrees as expected and the piston moves forward and backward for a complete cycle, similar to that of Figure 7-7, since the simulation duration is *1* second.

Also note that the *Mates* node in the *MotionManager* tree, where the number of redundancies is *7* (see Figure 7-8), as expected. Right click the *Mates* node and choose *Degrees of Freedom* to show details of the redundancies (see Figure 7-9), in which the redundant degrees of freedom removed by *Motion* are listed. Note that no reaction forces are recorded for the joints at the redundant degree of freedom when a dynamic simulation is carried out.

If you encounter a problem in conducting a motion analysis, try to run a *Basic Motion* analysis first. Occasionally, *Motion Analysis* may have problems in addressing redundant constraints; *Basic Motion* analysis seems to clear up the problem.

Saving and Reviewing Results

We will create four graphs for the mechanism: *X*-position, *X*-velocity, and *X*-acceleration at the mass center of the piston, and angular velocity of mate *Concentric2* (between crank and rod).

Right click *piston<1>* in the *MotionManager* tree, and choose *Create Motion Plot*. In the *Results* dialog box, choose *Displacement/Velocity/Acceleration*, select *Center of Mass Position*, and then *X Component*.

Figure 7-9

A graph will appear similar to that of Figure 7-10. Note that from the graph, the piston mass center moves between about *5* and *11* in. horizontally, in reference to the global coordinate system, in which the origin of the coordinate system coincides with the center point of the hole in the bearing.

At the starting point, the crank is at the upright position, and the piston is located at *7.42* in. (that is, $\sqrt{8^2 - 3^2}$) to the right of the origin of the global coordinate system. Note that the lengths of the crank and rod are *3* and *8* in., respectively. When the crank rotates to *90* degrees counterclockwise, the position becomes *5* (which is *8–3=5*). When the crank rotates *270* degrees, the piston position is *11* (which is *8+3*).

Repeat the same steps to create graphs for the velocity (choose *Linear Velocity*) and acceleration (choose *Linear Acceleration*) of the piston mass center in the *X*-direction. The graphs should be similar to those of Figures 7-11 and 7-12, respectively.

Figure 7-10 *X*-Position of the Piston

Figure 7-11 *X*-Velocity of the Piston

Figure 7-12 *X*-Acceleration of the Piston   Figure 7-13 Angular Velocity of *Concentric2*

The graph of the angular velocity of the mate *Concentric2 (crank<1>, rodandpin<1>)* can be created by right clicking the mate in the *MotionManager* tree, choosing *Creating Motion Plot*, and selecting *Displacement/Velocity/Acceleration, Angular Velocity*, and then *Z Component* in the *Results* dialog box. The graph should be similar to that of Figure 7-13.

If you expand the *Results* node in the *MotionManager* tree, you should see there are four graphs listed, *Plot1<Center of Mass Position1>, Plot2<Velocity1>, Plot3<Acceleration1>*, and *Plot4<Angular Velocity1>*, as shown in Figure 7-14.

Interference Check

Next we learn how to perform interference checking. *Motion* allows you to check for interference in your mechanism as constituent components move. In addition to the interference checking in *Motion*, you can detect interference between components in the *SOLIDWORKS* assembly model, regardless of whether a component participates in the motion model.

Using the interference checking capability in *Motion*, you can find the interference that occurs between the selected components as the mechanism moves through a specified duration of motion.

Figure 7-14

Make sure you have completed a simulation before proceeding to the interference check. Right click the root assembly *Lesson7* in the *MotionManager* tree and then select *Check Interference*. The *Find Interferences Over Time* dialog box appears (Figure 7-15).

Pick *piston<1>* and *rod<1>* (expanding *rodandpin<1>*) from the *MotionManager* tree to be included in the interference check. The *Start Frame, End Frame*, and *Increment* allow you to specify the motion frame used as the starting position, final position, and increment in between for the interference check. We will enter *1, 201*, and *1*, for the *Start Frame, End Frame*, and *Increment*, respectively. Click the *Find Now* button (circled in Figure 7-15) to start the interference check.

After pressing the *Find Now* button, the mechanism starts moving, and the crank rotates a complete cycle. At the same time, the list in the *Find Interferences Over Time* dialog box continuously grows. The list at the lower half of the dialog box shows all interference conditions detected. The frame, simulation time, parts that caused the interference, and the volume of the interference detected are listed.

Select the *Index* number of an entity (for example, *5*, as shown in Figure 7-16) to enable the *Details* button and the *Magnifying* button at the lower right corner. Selecting *Details* displays more information about the interference. Click the *Magnifying* button to zoom in to the location of the interference. As shown in Figure 7-17, the interference occurs between the inner face of the piston and the outer end face of the rod. You may want to choose the *Front* view to see the interference (Figure 7-18).

Note that you may see a different result than those of Figures 7-16 to 7-18 if you choose subassembly *rodandpin<1>* instead of *rod<1>*. The interference checking capability of *Motion 2024 SP1.0* is not able to capture interference correctly for subassemblies due to most likely a software bug. The software bug was reported to SOLIDWORKS company on 1/19/2010 (SR 1-1495511201). Therefore, make sure you expand the subassembly *rodandpin<1>* and then choose *rod<1>* part.

Figure 7-15 The *Find Interference Over Time* Dialog Box

Figure 7-16 Interference Check Results

Figure 7-17 Interference Identified

If you want to check the interference between components that involve one or more subassemblies, a workaround is to use the interference detection capability in the assembly mode. You may adjust the view of the motion model (for example, like that of Figure 7-18) and choose *Tools > Evaluate > Interference Detection* to find the interference between a part and a subassembly or for the entire mechanism.

Close the dialog box and save the model.

Figure 7-18 Interference detected (Front View)

Creating and Running a Dynamic Analysis

Duplicate the motion model and rename it *Dynamic*.

A force simulating the engine firing load (acting along the negative *X*-direction) will be added to the piston for a dynamic simulation. It will be more realistic if the force can be applied just before the piston starts moving to the left (negative *X*-direction) and if the force can be applied only for a selected short period of time.

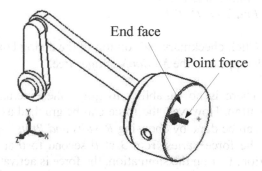

Figure 7-19 Pick the End Face for Force Application

In order to do so, we will have to define measures that monitor the position of the piston for the firing load to be activated. Such a motion study can be created by using event-based motion study. We will introduce the event-based motion study in *Lesson 8*.

In this lesson, we simply create a force as a step function of *3* lb$_f$ along the negative *X*-direction and apply it for *0.1* seconds. The force will be defined as a point force at the center point of the end face of the piston, as shown in Figure 7-19.

Before adding the force, we will suppress the rotary motor. From the *MotionManager* tree, right click *RotaryMotor1*, and choose *Suppress*. The motor will be turned off. Note that if you run a simulation now, nothing will happen since there is no motion driver or force defined (gravity has not been added).

Now we are ready to add the force. Click the *Force* button ↖ from the *Motion* toolbar. In the *Force* dialog box (Figure 7-20), choose *Linear Force* (default) for *Force Type* and *Action Only* (default) under *Direction*, and then pick the end face of the piston, as shown in Figure 7-19 for the location of the applied force.

Figure 7-20

The *Face<1>@piston-1* is now listed in the dialog box (Figure 7-20), and a force arrow will appear on the end face of the piston in the graphics area, indicating that the force will be applied to the center of the end face and in the direction that is normal to the face, i.e., in the positive *X*-direction. Click the small arrow button ⤴ (circled in Figure 7-20) to flip the force direction. The force arrow in the graphics area should now be reversed like that of Figure 7-19. That is, the force will be applied to the center of the end face of the piston and in the negative *X*-direction.

Scroll down the box to define *Force Function*. Click the *Function* tab, choose *Step* function (see Figure 7-20), and enter the following:

> *Initial Value F1: 3*
> *Start Time t1: 0*
> *Final Value F2: 0*
> *End Time t2: 0.1*

Click checkmark ✔ on top of the dialog box to accept the definition. A force node (name: *Force1*) should appear in the *MotionManager* tree.

There is no capability to graph the external force defined as a step function prior to a motion simulation. However, the force can be graphed as a reaction force after completing a motion simulation. This can be done by choosing *Results and Plots* 📈 from the *Motion* toolbar, as shown in Figure 7-21. In fact, the force varies from *3* at *0* second to *0* at *0.1* seconds as a smooth curve instead of a true step function. During the simulation, the force is activated at the beginning, i.e., *0* second.

Before running a simulation, we will increase the number of frames in order to see more refined results and graphs. Click the *Motion Study Properties* button ⚙ from the *Motion* toolbar. In the *Motion Study Properties* dialog box, enter *500* for *Frames per second*, and click the checkmark on top.

Click *Calculate* button 📊 from the *Motion* toolbar to simulate the motion. A one-second simulation will be carried out. After a short moment, you should see the mechanism starts moving. The mechanism will move and the crank will make several turns before reaching the end of the simulation duration.

Figure 7-21

Graphs created in the kinematic simulation are available for displaying dynamic simulation results, such as the piston position, etc. As shown in Figure 7-22, the piston moves along the negative *X*-direction for about *0.15* seconds before reversing its direction.

It is also evident in the velocity graph shown in Figure 7-23 that the velocity changes sign at three instances (close to *0.15*, *0.48*, and *0.78* seconds). Recall that the force was applied for the first *0.1* seconds. Had the force application lasted longer, the piston could be continuously pushed to the left (negative *X*-direction) even when the piston reaches the left end and tries to move to the right (due to inertia). As a result, the crank would have been oscillating at the left of the center of the bearing, i.e., between *0* and *180* degrees about the *Z*-axis, without making a complete turn.

Figure 7-22 *X*-Position of the Piston

Figure 7-23 *X*-Velocity of the Piston

Next, we graph the reaction force on the piston due to the applied force. The reaction force can be graphed only at assembly mates that constrain certain degrees of freedom. As discussed in *Lesson 1*, these reaction forces at mates (or joints) are keeping the moving bodies connected following Newton's 3rd Law. Graphing the reaction force in *Motion* is a little tricky due to the redundant mates assigned by the *Motion* solver internally. When graphing the reaction forces, you may see zero forces or moments if redundant degrees of freedom of a mate are selected. If this happens, you may want to pick another mate that constrains the same degree of freedom to graph the non-zero reaction forces.

In this example, we will create a plot by selecting the piston and then picking *Concentric3* that connects the pin to the piston to graph the reaction force.

From the *MotionManager* tree, right-click *piston<1>* and choose *Create Motion Plot*. In the *Results* dialog box, choose *Forces*, select *Reaction Force*, and *X Component*. Then, clear the face listed and pick *Concentric3* (from *MotionManager* tree), as shown in Figure 7-24. Click the checkmark to accept the definition.

Figure 7-24

Figure 7-25 *X*-Reaction Force of *Concentric3*

A warning message box appears, indicating that the motion model has redundant constraint which can lead to invalid results, and asking you to replace redundant constraints with bushings. Select *No* to move on. A graph like that of Figure 7-25 should appear. There are three peaks in Figure 7-25 representing the time when the largest reaction forces occur at the joint, which occurs at close to *0.15*, *0.48*, and *0.78* seconds, i.e., when the piston reverses its moving direction. The results seem to make sense. Note that in order to capture the peak force, you will need to increase *Frames per Second* for motion analysis. In this example, 500 frames per second seems to be adequate.

Save the model. We will carry out theoretical calculations to verify the simulation results next. We will focus on kinematic analysis.

7.4 Result Verifications

In this section, we verify the motion analysis results using kinematic analysis theory often found in mechanism design textbooks. Note that in kinematic analysis, position, velocity, and acceleration of given points or axes in the mechanism are analyzed.

In kinematic analysis, forces and torques are not involved. All bodies (or links) are assumed massless. Hence, mass properties defined for bodies are not influencing the analysis results.

The slider-crank mechanism is a planar kinematic analysis problem. A vector plot that represents the positions of joints of the planar mechanism is shown in Figure 7-26. The vector plot serves as the first step in computing position, velocity, and accelerations of the mechanism.

The position equations of the system can be described by the following vector summation,

$$Z_1 + Z_2 = Z_3 \qquad\qquad (7.1)$$

where

$Z_1 = Z_1 \cos \theta_A + i Z_1 \sin \theta_A = Z_1 e^{i\theta_A}$

$Z_2 = Z_2 \cos \theta_B + i Z_2 \sin \theta_B = Z_2 e^{i\theta_B}$

$Z_3 = Z_3$, since θ_C is always 0.

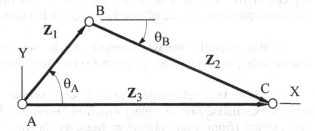

Figure 7-26 Vector Plot of the
Slider-Crank Mechanism

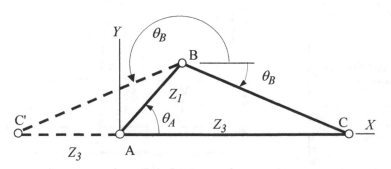

Figure 7-27 Two Possible Configurations

The real and imaginary parts of Eq. 7.1, corresponding to the X and Y components of the vectors, can be written as

$$Z_1 \cos \theta_A + Z_2 \cos \theta_B = Z_3 \tag{7.2a}$$

$$Z_1 \sin \theta_A + Z_2 \sin \theta_B = 0 \tag{7.2b}$$

In Eqs. 7.2a and 7.2b, Z_1, Z_2, and θ_A are given. We are solving for Z_3 and θ_B. Equations 7.2a and 7.2b are non-linear. Solving them directly for Z_3 and θ_B is not straightforward. Instead, we will calculate Z_3 first, using trigonometric relations, i.e.,

$$Z_2^2 = Z_1^2 + Z_3^2 - 2Z_1 Z_3 \cos \theta_A$$

Hence,

$$Z_3^2 - 2Z_1 \cos \theta_A Z_3 + Z_1^2 - Z_2^2 = 0$$

Solving Z_3 from the above quadratic equation, we have

$$Z_3 = \frac{2Z_1 \cos \theta_A \pm \sqrt{(2Z_1 \cos \theta_A)^2 - 4(Z_1^2 - Z_2^2)}}{2} \tag{7.3}$$

where two solutions of Z_3 represent the two possible configurations of the mechanism shown in Figure 7-27. Note that point C can be either at C or C' for any given Z_1 and θ_A.

From Eq. 7.2b, θ_B can be solved by

$$\theta_B = \sin^{-1}\left(\frac{-Z_1 \sin \theta_A}{Z_2}\right) \tag{7.4}$$

Similarly, θ_B has two possible solutions corresponding to vector $\mathbf{Z}_3$.

Taking derivatives of Eqs. 7.2a and 7.2b with respect to time, we have

$$-Z_1 \sin \theta_A \, \dot\theta_A - Z_2 \sin \theta_B \, \dot\theta_B = \dot{Z}_3 \tag{7.5a}$$

$$Z_1 \cos \theta_A \, \dot\theta_A + Z_2 \cos \theta_B \, \dot\theta_B = 0 \tag{7.5b}$$

where $\dot\theta_A = \dfrac{d\theta_A}{dt} = \omega_A$ is the angular velocity of the rotary motor, which is a constant. Note that Eqs.

7.5a and 7.5b are linear functions of $\dot{Z}_3$ and $\dot\theta_B$. Rewrite the equations in a matrix from

$$
\begin{bmatrix} Z_2 \sin\theta_B & 1 \\ Z_2 \cos\theta_B & 0 \end{bmatrix} \begin{bmatrix} \dot\theta_B \\ \dot Z_3 \end{bmatrix} = \begin{bmatrix} -Z_1 \sin\theta_A \, \dot\theta_A \\ -Z_1 \cos\theta_A \, \dot\theta_A \end{bmatrix} \tag{7.6}
$$

Equation 7.6 can be solved by

$$
\begin{bmatrix} \dot\theta_B \\ \dot Z_3 \end{bmatrix} = \begin{bmatrix} Z_2 \sin\theta_B & 1 \\ Z_2 \cos\theta_B & 0 \end{bmatrix}^{-1} \begin{bmatrix} -Z_1 \sin\theta_A \, \dot\theta_A \\ -Z_1 \cos\theta_A \, \dot\theta_A \end{bmatrix}
$$

$$
= \frac{1}{-Z_2 \cos\theta_B} \begin{bmatrix} 0 & -1 \\ -Z_2 \cos\theta_B & Z_2 \sin\theta_B \end{bmatrix} \begin{bmatrix} -Z_1 \sin\theta_A \, \dot\theta_A \\ -Z_1 \cos\theta_A \, \dot\theta_A \end{bmatrix}
$$

$$
= \frac{1}{-Z_2 \cos\theta_B} \begin{bmatrix} Z_1 \cos\theta_A \, \dot\theta_A \\ Z_1 Z_2 \cos\theta_B \sin\theta_A \, \dot\theta_A - Z_1 Z_2 \sin\theta_B \cos\theta_A \, \dot\theta_A \end{bmatrix}
$$

$$
= \begin{bmatrix} -\dfrac{Z_1 \cos\theta_A \, \dot\theta_A}{Z_2 \cos\theta_B} \\[4mm] -\dfrac{Z_1 \left(\cos\theta_B \sin\theta_A \, \dot\theta_A - \sin\theta_B \cos\theta_A \, \dot\theta_A \right)}{\cos\theta_B} \end{bmatrix} \tag{7.7}
$$

Hence

$$
\dot\theta_B = -\frac{Z_1 \cos\theta_A \, \dot\theta_A}{Z_2 \cos\theta_B} \tag{7.8}
$$

and

$$
\dot Z_3 = Z_1 \left(\tan\theta_B \cos\theta_A \, \dot\theta_A - \sin\theta_A \, \dot\theta_A \right) \tag{7.9}
$$

In this example, $Z_1 = 3$, $Z_2 = 8$, and the initial conditions are $\theta_A(0) = \pi/2$ and $\theta_B(0) = \sin^{-1}(3/8)$. Note that these initial conditions are different from those of *Motion* model, where $\theta_A(0) = 0$. You may graph the angular displacement of the crank (see Figure 7-28).

The angular displacement starts at zero (when the crank is in the upright position; that is $\theta_A(0) = \pi/2$), increases to *180* degrees (crank rotates 180 degrees, vertically downward), drops to *–180* degrees (with the same physical configuration), and comes back to zero. The discontinuous angular displacement shown in *Motion* is artificial, resulting from the way that the angular displacement was measured in *Motion*. The solutions can be implemented using a spreadsheet. The *Excel* spreadsheet file, *lesson7.xls*, can be found at the publisher's web site.

As shown in Figure 7-29, Columns A to I represent time, Z_1, Z_2, $\dot{\theta}_A$, θ_A, Z_3, θ_B, $\dot{Z}_3$, and $\dot{\theta}_B$, respectively. Note that in this calculation, $Z_3(0)>0$ is assumed, hence $\theta_B(0)<0$ (clockwise), as illustrated in Figure 7-27. This is consistent with the initial configuration we created in the *Motion* model.

Figures 7-30 and 7-31 show the graphs of data in Columns F and H, respectively. Comparing Figures 7-30 and 7-31 with Figures 7-10 and 7-11, the simulation analysis results are verified.

Note that the accelerations of a given joint in the mechanism can be formulated by taking one more derivative of Eqs. 7.5a and 7.5b with respect to time. The resulting two coupled equations can be solved using *Excel* spreadsheet. This is left as an exercise.

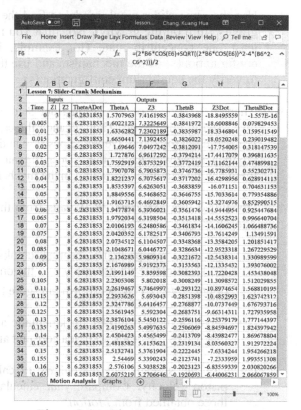

Figure 7-29 The *Excel* Spreadsheet

Figure 7-28

Figure 7-30 Position of the Piston (Column F)

Figure 7-31 Velocity of the Piston (Column H)

Exercises:

1. Derive the acceleration equations for the slider-crank mechanism by taking derivatives of Eqs. 7.5a and 7.5b with respect to time. Solve these equations for the linear acceleration of the piston and the angular acceleration of the mate *Concentric2*, using a spreadsheet. Compare your solutions with those obtained from *Motion*.

2. Use the same slider-crank model to find the equilibrium configuration(s) of the mechanism due to gravity (turn on the gravity). Note that you need to turn off the rotary motor, and add friction to at least one concentric mate (for example, between bearing and crank). Show the equilibrium configuration(s) of the mechanism and use the energy method you learned from Sophomore *Statics* to verify the equilibrium configuration(s).

3. Change the length of the crank from *3* to *5* in. in *SOLIDWORKS*. Repeat the kinematic analysis discussed in this lesson. Then, change the crank length from *3* to *5* in. in the spreadsheet (*Microsoft Excel* file, *lesson7.xls*). Generate position and velocity graphs from both *Motion* and the spreadsheet. Do they agree with each other? Does the maximum slider velocity increase due to a longer crank? Is there any interference occurring in the mechanism?

4. Download five *SOLIDWORKS* parts from the publisher's web site to your computer (folder name: *Exercise 7-4*).

    (i)   Use these five parts, i.e., bearing, crankshaft, connecting rod, piston pin, and piston (see Figure E7-1), to create an assembly like the one shown in Figure E7-2. Note that the crankshaft must orient at *45°* CCW, as shown in Figure E7-2.

    (ii) Create a motion model for kinematic analysis. Conduct motion analysis by defining a driver that drives the crankshaft at a constant angular speed of *1,000* rpm.

    (iii) Use the spreadsheet *lesson7.xls* to calculate the piston velocity. Compare your calculations with those obtained from *Motion*.

Figure E7-1 Five *SOLIDWORKS* Parts   Figure E7-2 Assembled Configuration

Lesson 8: A Slider-Crank Mechanism — Event-Based Dynamic Simulation

8.1 Overview of the Lesson

In *Lesson 7*, we created an engine firing force as a step function of 3 lb$_f$ in 0.1 second applied to the exterior end face of the piston at the beginning of the simulation. Such a force did not simulate a realistic motion of an internal combustion engine. We stated that a firing load would be more realistic if the force was applied when the piston started moving to the left (negative *X*-direction) and was applied only for a prescribed short period of time. Such a desired firing load can be created by using event-based motion study capability, which is offered in *SOLIDWORKS Simulation® Professional* added in.

Motion studies can be time-based or event-based. Time-based motion studies describe the response to time-based changes in motion elements on the assembly motion. All lessons we discussed so far are time-based. Event-based motion studies are defined with a set of motion actions resulting from triggering events. You may create an event-based motion study when you do not know the exact time sequence for changes that alter the motion of the mechanism. You can obtain the time sequence for the changes by calculating an event-based motion study.

In this lesson, we revisit the slider-crank example and create a motion study that simulates the motion of an internal combustion engine with a more realistic firing load using event-based motion analysis. We start by offering a short discussion on the event-based motion analysis capability, then we discuss the slider-crank assembly with an artificially added solid part (serves as a timer) that helps detect the triggering event. We discuss the motion model to be created before starting *SOLIDWORKS* and going over the example. You will learn the concept of event-based motion analysis and implementation of such a simulation in detail.

8.2 Event-Based Motion Analysis

Event-based motion requires a set of tasks. The tasks can be sequential or can overlap in time. Each task is defined by a triggering event and its associated task action. Task actions control or define motion during the task.

A task trigger is the event that drives the motion control action for a task. You can define task triggers based on time, previous tasks, or sensed values, such as part position in an assembly.

A task action defines or constrains the motion of one or more parts in the assembly. You can define actions to suppress or activate mates, stop motion, or to change values for motors, forces, or torques.

In this example, we position a timer, which is a small round solid bar, at *X = 2.5* in., and Y = *0.758* in. from the center of the bearing (see Figure 8-1a). The crank interferes with the timer when it rotates about 6° from its initial position, in which it aligns horizontally with the connecting rod, as illustrated in Figure 8-1b. Crank colliding with the timer is defined as a task trigger. When the collision occurs, i.e., the crank rotates about $6°$ CCW from the *X*-axis, the firing force will be activated for a short period of time (in this case 0.01 seconds), which is defined as the task action.

(a) Position of the Timer   (b) Crank Colliding with the Timer at $6°$ CCW

Figure 8-1 Crank Colliding with the Timer

8.3 The Slider-Crank Example

SOLIDWORKS Part and Assembly Files

We start this lesson with the same parts and subassembly as those of *Lesson 7*, plus one more part called *timer*. As discussed, the part *timer* is a small round solid bar assembled to the slider-crank assembly at $X = 2.5$ in., and $Y = 0.758$ in. from the center of the bearing (see Figure 8-1a). There are three assembly mates created to assemble the *timer*: *Coincident6*, *Distance1*, and *Distance2*, as shown in Figure 8-2.

Another setup different from *Lesson 7* is the initial orientation of the crank. Instead of an upright position, the crank is positioned in a horizontal direction, aligning with the rod and the piston. This was achieved by replacing the coincident mate (*Coincident2* shown in Figure 7-3a) with an angle mate (*Angle1*), in which the *Right Plane* of the crank is perpendicular to the *Right Plane* of the assembly (see Figure 8-3).

All parts and subassembly are properly assembled in the assembly model prepared for this lesson. We will start the lesson by opening the assembly file *Lesson8.SLDASM*. In addition, the file *Lesson8withResults.SLDASM* contains the complete motion simulation model and results for your references.

Simulation Model

Once we suppress the mate *Angle1*, the mechanism has one free degree of freedom, either rotating the crank or the rod (about the axis of the concentric mates, i.e., Z-axis), or moving the piston horizontally (along the *X*-direction). Either rotation or translation will be sufficient to uniquely determine the position, velocity, and acceleration of all parts in the mechanism. An initial angular velocity is added to the crank so that it rotates at *60* RPM CCW at the beginning. A constant force of *100* lbf is added to the center point of the end face of the piston, as shown in Figure 8-4. Also, a proximity sensor that detects the interference between the crank and the timer is added to the simulation model. In *Motion*, a proximity sensor monitors the assembly for interferences between a geometric entity (a line or a surface) you define and components you select.

Note that an interference sensor, also available in *Motion*, would have been a more logical choice for monitoring the interference between the crank and the timer. However, if you define an interference sensor in *SOLIDWORKS 2024 SP1.0*, the interference is not detected. This was a software bug and was reported on 4/2/2019 (SR 1-20935978046).

(a) Assembly Mate: *Coincident6*   (b) Assembly Mate: *Distance1*

(c) Assembly Mate: *Distance2*

Figure 8-2 Assembly Mates Defined for the Part: *timer*

Figure 8-3 *Angle1* Mate Defining Initial Configuration

Figure 8-4 Firing Force Applied at the End Face of the Piston

A working motion model with an interference sensor saved from *SOLIDWORKS Motion 2018 SP5.0* version can be found in the folder: *Lesson 8 Saved from 2018 Version*. You may open this model and play the motion of the *Condition Check* study. To preserve the results saved from the 2018 version, please do not click the *Calculate* button .

In this lesson, we discuss two motion studies: *Condition Check* and *Event-Based Dynamic Simulation*. The *Condition Check* motion study simply detects the interference between the crank and timer, and then ends the simulation. We graph the crank rotation angle to verify the angle at the interference to be about 6°.

The second study, *Event-Based Dynamic Simulation*, consists of three tasks. Task 1 turns off the firing force to eliminate the possibility of unintended force activation. Task 2 detects the interference. Task 3 activates the firing force of *100* lb$_f$ for *0.01* seconds once Task 2 starts, then turn the firing force off.

8.4   Using *SOLIDWORKS Motion*

Start *SOLIDWORKS* and open assembly file *Lesson8.SLDASM*.

First, we suppress the third assembly mate *Angle1*. From the *FeatureManager* tree, expand the *Mates* node, right click *Angle1*, and choose *Suppress* . The mate *Angle1* becomes inactive.

Before creating any entities, always check the units system. Make sure *IPS* units system is chosen for this example.

Adding a Sensor

From the *FeatureManager* tree (*SOLIDWORKS* browser), right click *Sensors* and choose *Add Sensor*.

In the *Sensor* dialog box (see Figure 8-5), choose *Proximity* for *Sensor Type*, pick *timer* for *Proximity sensor location* and *crank* for *Components to track* from the graphics area, enter *1*in. for *Proximity sensor range*, select *Alert*, and choose *is true* for *Notify me if the value*. Click the checkmark on top of the dialog box to accept the sensor definition. An entity, *Proximity1*, is added to *Sensors* in the *FeatureManager* tree (see Figure 8-6).

Note that the value of the sensor indicates whether or not interferences are detected between the sensing line we define (added to the *timer* by *Motion*, as shown in the zoom-in view of the *timer* in Figure 8-5) and the crank you select. The length of the sensing line is 1in., which is sufficiently large to detect interference with the crank.

Entering Motion for Motion Study 1: Condition Check

Click the *Motion Study* tab at the bottom of the graphics area to bring up the *MotionManager* window. Double click the tab and rename the study *Condition Check*. Choose *Motion Analysis* for the study (default selection could be *Animation*, which will not be adequate).

Defining Initial Velocity

We add an initial angular velocity of *60* RPM CCW to the crank. Click *crank* from the *MotionManager* tree, press the right mouse button, and choose *Initial Velocity*; the *Initial Velocity* dialog box will appear (Figure 8-7).

Click the field under *Initial Angular Velocity*, and from the graphics area, pick a circular arc of the crank part. A small arc with an arrow appears. Click the direction button 🔁 for a CCW direction if necessary. Enter *60* RPM, as shown in Figure 8-7. Click the checkmark ✔ to accept the initial velocity.

Pick the timer and crank

The line defined for the proximity sensor

Figure 8-5 Defining a Proximity Sensor to Detect Interference

Figure 8-6

Pick a circular arc of the crank

Figure 8-7 Defining Initial Angular Velocity

Defining Event-Based Motion Study

From the *Motion* tool bar, click the *Event-based Motion View* button ▦ (circled in Figure 8-8) or right click the root node *Lesson 8* and select *Event-based Motion View* to bring up the event window (see Figure 8-8).

Note that if you do not see the event view like that shown in Figure 8-8, you may need to consult with your system administrators to make sure your licenses include either *Simulation Professional* or *Simulation Premium*.

Click the *Click here to add* button ✚ *Click here to add* (circled in Figure 8-8) to add a task. A default *Task1* is added, as shown in Figure 8-9.

Click the *Select* button ⬚ under *Trigger* (circled in Figure 8-9). In the *Trigger* dialog box (Figure 8-10), expand *Sensors* and select *Proximity1*. Click *OK* to accept the selection.

Alert On is added under *Condition* of *Triggers*, and *End Motion* is added under *Feature* of *Actions*, as circled in Figure 8-11. This task implies that when the interference is detected, end the motion analysis.

Defining and Running Simulation

Choose *Motion Analysis* for study. Click the *Motion Study Properties* button ⚙ from the *Motion* toolbar. In the *Motion Study Properties* dialog box, enter *600* for *Frames per second*, and click the checkmark on top.

Figure 8-8   The Event Window

Figure 8-9 Defining *Task1*

Figure 8-11 *Task1* Completely Defined

Figure 8-10

Click the *Timeline View* button to bring the timeline window back. Drag the end time key to *0.1* second mark in the timeline area to define the simulation duration. Zoom in to the timeline area if needed. Make sure that the *Angle1* mate is suppressed.

Click the *Calculate* button from the *Motion* toolbar to simulate the motion. A *0.1*-second simulation will be carried out. After a short moment, you should see that the mechanism starts moving. The crank rotates about *6* degrees as expected and the simulation stops.

Reviewing Results

We will create a graph to display the angular position of the crank.

From *MotionManager* tree, right click *crank*, and choose *Create Motion Plot*. In the *Results* dialog box, choose *Displacement/Velocity/Acceleration*, select *Angular Displacement*, and then *Magnitude*. Click the checkmark to accept the graph. The graph will appear similar to that of Figure 8-12. Note that from the graph, the crank rotates from *–90* to *–83.4* degrees in about *0.019* seconds, making about a *6.6* degree CCW turn as expected.

We completed the first motion study. Save your model before moving to the next motion study.

Motion Study 2: Event-Based Dynamic Simulation

Duplicate a motion study by right clicking *Condition Check* and select *Copy Study*. Rename the study as *Event-Based Dynamic Simulation*. Delete *Task 1* created in the *Condition Check* study. In this study, we create three new tasks.

Task 1 turns off the firing force to make sure the force is not activated unless we tell it so.

Task 2 detects the interference, serving as the event trigger for Task 3.

In Task 3, we start the firing force of *100* lb_f for *0.01* seconds as soon as Task 2 is on, i.e., the interference between the crank and timer is detected. The firing force is turned off after *0.01* seconds.

Figure 8-12

Figure 8-13

Figure 8-14

Adding a Force

Click the *Force* button ⬉ from the *Motion* toolbar. In the *Force/Torque* dialog box (Figure 8-13), choose *Action Only* (default) under *Direction*, and then pick the exterior end face of the piston, as shown in Figure 8-4 for the location of the applied force.

The *Face<1>@piston-1* is now listed in the dialog box (circled in Figure 8-13), and a force arrow will appear on the end face of the piston in the graphics area, indicating that the force will be applied to the center of the end face and in the direction that is normal to the face, i.e., in the positive *X*-direction. Click the small arrow button ↗ to flip the force direction. The force arrow in the graphics area should now be reversed like that of Figure 8-4. That is, the force will be applied to the center of the end face of the piston and in the negative *X*-direction.

Choose *Constant* and enter *100* lbf. Click the checkmark ✔ to accept the force. A force node (name: *Force1*) should appear in the *MotionManager* tree.

Defining Tasks

As discussed, we will define three tasks.

Same as before, we click the *Event-based Motion View* button ▦ (circled in Figure 8-8) to bring up the event window. Click the *Click here to add* button ➕ *Click here to add* (circled in Figure 8-8) to add a task. A new task is added. Rename the task *Task1*.

Click the *Select* button ⬚ under *Trigger*. In the *Trigger* dialog box (Figure 8-14), select *Time*. Click *OK* to accept the selection.

Enter *0* under *Time/Delay* of *Triggers*. Click the *Select* button ⬚ under *Feature* of *Actions*. In the *Feature* dialog box (Figure 8-15), expand *Forces* and select *Force1*. Click *OK* to accept the selection.

Choose *Off* under *Action* of *Actions*.

We add *Task 2* and enter/select the followings:

Triggers/Trigger: *Proximity1*
Triggers/Condition: *Alert On*
Triggers/Time/Delay: *None*
Actions/Feature: *Force1*
Actions/Action: *On*

Task 2 does not activate the force but simply turns alert on when the interference is detected, or *Proximity1* sensor value becomes *is true*.

Similarly, we add *Task 3* and enter/select the followings:

Task 3:
Triggers/Trigger: *Task2*
Triggers/Condition: *Task Start*
Triggers/Time/Delay: *0.01s delay*
Actions/Feature: *Force1*

Figure 8-15

Actions/Action: *Off*

Task 3 activates *Force1* for *0.01* seconds, and turns the force off after the *0.01s* delay.

A complete set of tasks defined is now listed in the *Event-based* area, as shown in Figure 8-16.

Tasks		Triggers			Actions					Time	
Name	Description	Trigger	Condition	Time/Delay	Feature	Action	Value	Duration	Profile	Start	End
Task1		Time		0s	Force1	Off					
Task2		Proximity1	Alert On	<None>	Force1	On					
Task3		Task2	Task St	0.01s delay	Force1	Off					
	+ Click here to add										

Figure 8-16 The Three Tasks Listed in the *Event-Based* Area

Defining and Running Simulation

Choose *Motion Analysis* for the study. Click the *Motion Study Properties* button ⚙ from the *Motion* toolbar. In the *Motion Study Properties* dialog box, enter *600* for *Frames per second*, and click the checkmark on top. Click the *Timeline View* button 📄 to bring the timeline window back. Drag the end time key to *0.35* second mark in the timeline area to define the simulation duration. Zoom in to the timeline area if needed.

Click the *Calculate* button 📊 from the *Motion* toolbar to simulate the motion. A *0.35*-second simulation will be carried out. After a short moment, you should see that the mechanism starts moving. The crank rotates slowly since *Motion* is checking interference at every single timeframe in the analysis.

Note again that if you chose to create an interference sensor (instead of a proximity sensor), the interference is not detected due to a software bug in *SOLIDWORKS 2024 SP1.0*. As a result, the crank rotates due to the initial rotation velocity. Force has never been applied.

Reviewing Results

From *Motion* toolbar, click the *Timeline View* button 📄 to bring up the timeline window (see Figure 8-17). Take a look at the time keys of *Force1*. There are nine pairs of time keys shown for *Force1*, indicating that the firing force was activated nine times. The first time key of individual pairs indicates the time when *Force1* is activated, which is when the interference between the crank and timer is first detected. You may drag the time line (the vertical line in the time area) to the location of the first key and check in the graphics area to see that the crank indeed interferes with the timer (see Figure 8-17). The second time key of the pair indicates that *Force1* is turned off after *0.01* seconds. The time intervals between the pairs that follow are shortened, indicating that the interference is detected sooner and sooner. That is, the crank rotates faster and faster due to the action of the firing load at every cycle as expected.

We also graph the force as results to see if the force is activated seven times with the correct magnitude and time interval.

We click *Results and Plots* button 📈 from the *Motion* toolbar, the *Results* dialog box appears (see Figure 8-18). Choose *Forces*, *Reaction Force*, and *X Component*. Click *Force1* from the *MotionManager* tree, and pick the bearing part from the graphics area as the reference part (that is, a reference part to which the data plotted in the result graph refers).

The result graph shown in Figure 8-18 indicates that the force is activated nine times for *0.01* seconds each time, which is correct.

The first time key indicating *Force1* is activated when interference is first detected

The second time key indicating *Force1* is turned off after *0.01* seconds

Figure 8-17 The Firing Force Activated 9 Times

Figure 8-18 The Graph of *Force1* in Results

Figure 8-19 Angular Position of the Crank

Figure 8-20 Angular Velocity of the Crank

Figure 8-21 *X*-Velocity of the Piston

Figure 8-22 *X*-Reaction Force of the Piston Pin
(*Concentric3*)

We will create four more graphs to display the angular position of the crank, angular velocity of the crank, *X*-velocity of the piston, and reaction force of the piston pin measured at *Concentric3*, as shown in Figure 8-19 to 8-22, respectively. The results all seem to make sense. One thing that raises a flag regarding design of structural integrity is the reaction force shown in Figure 8-22. At the end of the motion simulation (after the firing force is applied nine times), the reaction force acting on the piston pin increases to about *6,700* lbf. This reaction force seems to be excessive. The structural integrity of the piston pin must be verified by carrying out a structural analysis.

We completed the second motion study. Save your model for future references.

Exercises:

1. Use the same two parts as Exercise 5-1 to create an assembly like that of Figure E8-1. Create an event-based motion study to move the block for 300 mm along the X-direction, then turn the block to the left and move along the negative Z-direction for another 300 mm. Thereafter, turn the block to the right and move along the positive X-direction for another 100 mm.

Figure E8-1

Lesson 9: A Rail Carriage Example

9.1 Overview of the Lesson

In this lesson, we will learn an advanced feature in *SOLIDWORKS*, the *Path Mate*, which was first added to *SOLIDWORKS 2008*. Most assembly mates impose constraints to part motion between regular surfaces, such as flat surfaces and cylindrical surfaces. As a result, mating parts are allowed to translate or rotate along a prescribed direction.

In path mate, a vertex in a part is selected to move along a curve (or a loop composed of several curves) of the mating part. As a result, path mate allows a part to move along a curve slot, a groove, or fluting, varying its moving direction specified by the path curve. In addition, the pitch, yaw and roll of the moving part can be defined to resemble the physical conditions. Such a capability offered in *SOLIDWORKS* supports animation and motion analysis for a broader set of applications involving curvilinear motion.

In this lesson, we use the rail and the top portion of the carriage of a bathroom transporting device shown in Figure 9-1 as an example. This was an undergraduate design project carried out a few years ago for a physically challenged person.

This transporting device was created for the purpose of transporting a disabled woman from wheelchair to toilet and to shower bench without assistance. It also transports the person from the toilet or shower bench back to the wheelchair. The device is compact (to fit into a small bathroom), durable, and tailored to help the person to overcome her physical disability. The design features a three-button remote control that moves the person to the toilet, shower, and back to the wheelchair; a scissor lift with a linear actuator that provides lift; a carriage on a rail system that carries the person to designated locations; and a body support that safely holds the person while the system transports the person to designated locations. A second actuator mounted on top of the scissor lift provides a 90-degree rotation to the body support when the carriage is moved to a position right above the toilet so that the user will be properly oriented on top of the toilet. And a motor and a cable system are employed to pull the carriage.

Figure 9-1 The Bathroom Transporting

The design of the device was challenging since it has to accommodate the woman with disabilities, who can only use her right hand to operate the device. She will pull her wheelchair to the entrance of the bathroom, right in front of the device. She will use her right hand to move the two leg supports under her thighs, place the two arm supports under her arms, and press a button on the remote control mounted on top of the right arm support. The button pressing will trigger the actuator of the scissor lift to contract, creating a lift to pull her out of the wheelchair. Then, a motor will be activated to pull a cable that draws the carriage along the curve rail and transport her to the position above the toilet or shower bench. Position sensors are mounted on top of the rail to detect the location of the carriage and activate motor or actuators for desired motions.

While designing the device, path mates were employed to assemble the carriage to the rail, allowing the carriage to move along the rail. This motion animation supports the verification of the design concept, facilitates communications within the design team, and offers a demonstration of the design to the project sponsors and the user before physically building the device.

9.2 The Rail Carriage Example

The rail and carriage are important features of the device. The rail is a curve I-beam, created by sweeping an I-cross section sketch along an open loop curve composed of three straight lines and two circular arcs, as shown in Figure 9-2. The carriage consists of a base plate, two steerers, and four wheels, as shown in Figure 9-3. The wheels are sitting on the top faces of the bottom flange of the rail. A cable connecting to a motor is pulling the steerers to move the carriage along the rail. A universal joint under the base plate connects the scissor lift and the body support.

SOLIDWORKS Part and Assembly Files

Parts and assembly have been created in *SOLIDWORKS*. There are four parts and one subassembly: *Rail.SLDPRT*, *Base Plate.SLDPRT*, *Steerer.SLDPRT*, *Wheels.SLDPRT*, and *Carriage.SLDASM*. There are also two assembly model files, *Lesson9.SLDASM* and *Lesson9withResults.SLDASM*, created for this example.

Figure 9-2 The Rail and Carriage Subsystems

In this lesson, we will start with *Lesson9.SLDASM*, where only standard mates were employed. You will see the undesired motion of the carriage in this model and the deficiency of the standard mates employed. We will then replace the standard mates with path mates to create desired carriage motion. The assembly with path mates can be found in *Lesson9withResults.SLDASM*. You may want to open *Lesson9withResults.SLDASM* to see the desired motion of the carriage.

Figure 9-3 The Carriage Subassembly

The assembly *Lesson9.SLDASM* consists of one part and one subassembly: the rail (*Rail.SLDPRT*) and the carriage (*Carriage.SLDASM*). The rail is fixed to the ground, and the carriage is fully assembled to the rail by using five standard mates: three coincident and two tangent mates, as shown in Figure 9-4. You may want to expand the *Mates* node in the browser to see the list of mates. Move the mouse pointer to any of the mates; you should see the geometric entities selected for the mate highlighted in the graphics area.

The first mate (*Coincident1*) shown in Figure 9-4a aligns the two end faces of the rail and the steerer. This mate defines the initial position of the carriage, which will be suppressed to allow the carriage to move. The next two are tangent mates shown in Figure 9-4b, which constrain the outer cylindrical surface of the wheel (the two on the left side of the rail) to be tangent to the top face of the bottom flange of the rail. The final two are coincident mates, which place the mid-points of the axes in the respective steerers on the left surface of the rail web (see Figure 9-4c). In the carriage assembly, there are two sketches (*Sketch1* and *Sketch2*), where the two mid-points are defined (please see Figure 9-5). The carriage is fully assembled to the rail. You will not be able to move the carriage by dragging it in the graphics area.

Motion Model

In this example, the carriage will be given an initial velocity of *25* in./sec. along the negative Z-direction, as shown in Figure 9-6. The overall simulation is one second. A motion model of standard mates discussed above can be found in *Lesson9.SLDASM*.

(a) *Coincident1*   (b) *Tangent1* (and *Tangent2*)   (c) *Coincident2* (and *Coincident3*)

Figure 9-4 Standard Mates Defined for the Rail Carriage Example

Point13@Sketch1

Point9@Sketch2

Steerers

Figure 9-5 Mid-Points of axes in the Carriage

Initial velocity

Figure 9-6 The Motion Model

9.3 Using *SOLIDWORKS Motion*

Start *SOLIDWORKS* and open assembly file *Lesson9.SLDASM*. Click the *Motion Study* tab at the bottom of the graphics area to bring up the *MotionManager* window.

In the *MotionManager* tree, you should see that an initial velocity is listed, and a one-second simulation has been created. The first mate *Coincident1* should have been suppressed to allow the carriage to move. Click the *Play from start* ▐▶ button from the *Motion* toolbar to see the carriage motion. The carriage moves along the negative Z-direction. It moves out of the rail at the end of the simulation, as shown in Figure 9-7. This result is certainly undesirable. We will replace the standard mates with path mates for a desired carriage motion.

Defining Path Mates

First, we delete all mates except for *Coincident1*.

Figure 9-7 Carriage Moved Out of the Rail

Figure 9-8

Click the *Model* tab to go back to *SOLIDWORKS* assembly. From *SOLIDWORKS* browser, expand the *Mates* node. Click *Tangent1*, press the *Shift* key, and then click *Coincident3*. All four mates are highlighted like that of Figure 9-8. Right click and choose *Delete* to delete all four mates. Click the *Yes* button in the *Confirm Delete* dialog box four times (or choose *Yes to All*).

We will create two path mates. The first one will be defined by selecting *Point13* of *Sketch1* in the carriage and the sweep path of the rail (an open loop created in *Sketch2* of the rail part under feature *Sweep1*). Turn on the point display to see the points by choosing *View → Hide/Show → Points* from the pull-down menu. Do the same for sketch (*View → Hide/Show → Sketches*).

From *SOLIDWORKS* browser (see Figure 9-9), expand *Rail <1>*, then expand *Sweep1*, and choose *Sketch2*. You should see the sweep path highlighted in the graphics area (Figure 9-10a). This is the path we will pick for the path mate. Expand *Carriage<1>* and choose *Sketch1*. *Point13* will appear in the graphics area, as shown in Figure 9-10b. This is the vertex we will pick for the first path mate. Similarly, we will pick *Point9* of *Sketch2* to create the second path mate.

Insert a mate by choosing *Insert → Mate* from the pull-down menu or clicking the *Mate* button from the assembly toolbar. The *Mate* dialog box will appear (Figure 9-11). Choose *Advanced Mates* and click *Path Mate*. The *Component Vertex* of the *Mate Selections* is active and ready for you to pick a geometric entity. From the graphics area, pick *Point13*. Next, the *Path Selection* is active and ready for you to pick. We will first click *Selection Manager* button (right below the *Path Selection*), click *Select Open Loop* button in a set of buttons that appear in the graphics area, as shown in Figure 9-12, and then pick the beginning part of the sweep path. Right click to accept the open loop chosen. Both entities should now be highlighted in the graphics area and listed in *Mate Selections*. Click the checkmark to accept the path mate.

(a) Sweep Path   (b) Mid-Points

Figure 9-10 Rail Sweep Path and Mid-Points of Axes in the Carriage

Figure 9-9

For the second path mate, we will pick *Point9* and the same sweep path (make sure you choose open loop). Click the checkmark ✔ to accept the second mate. Click the checkmark one more time to close the dialog box. In the assembly browser, two path mates are added (*PathMate1* and *PathMate2*).

Pick the carriage in the graphics area and drag it. You should be able to drag the carriage through the path. However, the subassembly rolls when you move the carriage, which is not desirable. Choose *Undo* ↶ above the graphics area to restore the assembly to its initial configuration.

Next, we control the rolling motion of the carriage (see Figure 9-13 for pitch, yaw, and roll). We will create the *Roll Control* using the *Path Mate* dialog box (see Figure 9-11), where we will pick a vertical edge for the up vector, then select *Y* axis to align with the vector in order to constrain the rolling motion.

From the browser, right click *PathMate1*, and choose *Edit Feature* ⬚.

From the *PathMate1* dialog box, choose *Up Vector* for *Roll Control*, pick a vertical edge from the model (for example the vertical edge of the end face of the rail, as shown in Figure 9-14). The edge picked (*Edge<1>@Rail-1*) will appear in the dialog box, and a vertically downward arrow (that is, the up vector) will appear next to the edge picked.

Choose *Y* to align the *Y*-axis of the carriage (do not confuse it with the *Y*-axis of the global coordinate system, even though they are in the same direction in this example) to the up vector, as shown in Figure 9-14.

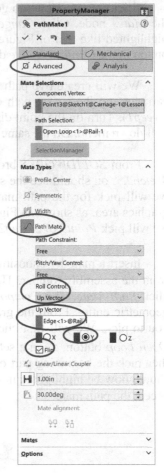

Figure 9-11

Pick the beginning of the sweep path

Select Open Loop button

Pick this point for *Component Vertex*

Figure 9-12 Pick Entities for the Path Mate

Roll

Yaw

Pitch

Figure 9-13 Roll, Yaw, and Pitch of the Carriage

If the carriage is positioned on top of the rail, click *Flip* to bring it to the right position, i.e., below the rail.

Click the checkmark ✅ on top of the dialog box to accept the modified mate definition. We will leave all the path, pitch, yaw, and roll control of the second path mate to *Free* since they have been constrained in the first path mate.

Now, pick the carriage in the graphics area and drag it. You should be able to drag the carriage through the path without rolling.

Choose *Undo* ↩ above the graphics area to restore the assembly to its initial configuration, or unsuppress *Coincident1* mate to bring the carriage back to its initial position. Suppress it to allow motion.

The up vector symbol

Pick a vertical edge

Y-axis of the carriage

Global coordinate system

Figure 9-14 Up Vector and Component *Y*-Axis

Save the model. We are ready to conduct a motion simulation.

Defining and Running Simulation

Click the *Motion Study* tab at the bottom of the graphics area. In the *MotionManager* window, choose *Motion Analysis* for study.

First, drag the end time key to the four-second mark (to allow the carriage to go through the entire path). We will refine the analysis time steps and choose a different solver (we will use *WSTIFF*).

Click the *Motion Study Properties* button ⚙ from the *Motion* toolbar. In the *Motion Study Properties* dialog box, enter *5000* for *Frame per second*, and click *Advanced Options* button (see Figure 9-15).

In the *Advanced Motion Analysis Options* dialog box (Figure 9-16), choose *WSTIFF*, and then click *OK*.

Note that *WSTIFF* and the default solver *GSTIFF* are similar in formulation and behavior. Both use a backwards difference formulation. They differ in that the *GSTIFF* coefficients are calculated assuming a constant step size, whereas *WSTIFF* coefficients are a function of the step size. If the step size changes suddenly during integration, *GSTIFF* introduces a small error, while *WSTIFF* can handle step size changes without loss of accuracy. Sudden step size changes occur whenever there are discontinuous forces, discontinuous motions or abrupt events, such as the carriage turning in this example.

Figure 9-15

Click the checkmark on top of the *Motion Study Properties* dialog box to accept the definition.

Click the *Calculate* button from the *Motion* toolbar to simulate the motion. A four-second simulation will be carried out. After a short moment, you should see that the carriage starts moving along the path like that of Figure 9-17. Note that if you see a warning dialog box, stating that:

The playback speed or frames per second settings for this motion study will result in poor performance ...

Simply click *No* to proceed since we intend to use a high frame rate for a more accurate solution.

This carriage motion is desirable. Be sure to save the model before exiting *SOLIDWORKS*.

Note that if a smaller *Frame per second* (see Figure 9-15) than 2,000 is entered for the study, the carriage stuck after making the first 90-degree turn at around 1.4 seconds. This was not the case for the 2009 year version or earlier.

Figure 9-16

Figure 9-17 Desired Carriage Motion

Exercises:

1. Create a motion model for the wheel to go through the curve groove of the ground shown in Figure E9-1. Note that the short axle of the wheel must align to the two small cuts of the groove horizontally, as shown in Figure E9-2.

Figure E9-2 Wheel Properly Aligned

Figure E9-1 The Wheel and Ground Parts

Notes:

Lesson 10: A Compound Spur Gear Train

10.1 Overview of the Lesson

In this lesson we learn how to simulate motion for a spur gear train system. A gear train is a set or system of gears arranged to transfer torque or energy from one part of a mechanical system to another. A gear train consists of driving gears that are mounted on an input shaft, driven gears mounted on an output shaft, and idler gears that interpose between the driving and driven gears in order to determine the direction of the output shaft and/or to increase the distance between the drive and driven gears. There are different kinds of gear trains, such as a simple gear train, compound gear train, epicyclic gear train, etc., depending on how the gears are shaped and arranged as well as the functions they intend to perform. The gear train we are simulating in this lesson is a compound gear train system, in which two or more gears are used to transmit torque or energy. All gears included in this lesson are spur gears; therefore, the shafts that these gears mounted on are in parallel.

In *Motion*, gear pair is defined as a *Mechanical Mate*, which couples two rotational motions of respective gear axes. Neither *SOLIDWORKS* nor *Motion* cares about the detailed geometry of the gear pair, i.e., if the gear teeth mesh adequately. You may even use cylinders or disks to represent the gears. No detailed tooth profile is necessary for any of the computations involving the gear mate. Apparently, force and moment between a pair of teeth in contact will not be calculated in gear train simulations. However, there are other important data being calculated by *Motion*, such as reaction force exerting on the driven shaft (for a dynamic analysis), which is critical for mechanism design. In any case, pitch circle diameters are essential for defining gear pairs and gear trains in *Motion*. The velocity ratio of the gear train, which is defined by the ratio of the angular velocities of the output and input gears, is determined by the pitch circle diameters of the individual gear pairs in the gear train system.

Although cylinders or disks are sufficient to model gears in *Motion*, we will use more realistic gear models in this lesson. All gears in the example are shown with detailed geometry, including teeth. In addition, detailed parts, including shafts, bearings, screws and aligning pins are included for a realistic gear train system, as shown in Figure 10-1. In this gear train simulation, we will focus more on graphical animation, less on computations of physical quantities. We will add a rotary motor to drive the input shaft.

10.2 The Spur Gear Train Example

Physical Model

The gear train example we are using for this lesson is part of a gearbox designed for an experimental lunar rover. The gear train is located in a gear box which is part of the transmission system of the rover, driven by a motor powered by solar energy. The purpose of the gear train is to convert a high-speed rotation and small torque generated by the motor to a low speed rotation and large torque output in order to drive the wheels of the rover. The gear train consists of four spur gears mounted on three parallel shafts, as shown in Figure 10-1.

Figure 10-1 The Gear Train System Example   Figure 10-2 Schematic View of the Gear Train

The four spur gears form two gear pairs: *Pinion 1* and *Gear 1*, and *Pinion 2* and *Gear 2*, as depicted in Figures 10-1 and 10-2. Note that *Pinion 1* is the driving gear that connects to the motion driver; e.g., a rotary motor. The motor rotates in a counterclockwise direction, therefore driving *Pinion 1*. *Gear 1* is the driven gear of the first gear pair, which is mounted on the same shaft as *Pinion 2*. Both *Gear 1* and *Pinion 2* rotate in a clockwise direction. *Gear 2* is driven by *Pinion 2*, and rotates in a counterclockwise direction. Note that the diameters of the pitch circles of the four gears are 50, 120, 60, and 125 mm, respectively; and the numbers of teeth are 25, 60, 24, and 50, respectively. Therefore, the circular pitch P_c, the diametral pitch P_d, and module m of the first gear pair are, respectively

$$p_c = \frac{\pi d_{p1}}{N_{p1}} = \frac{\pi d_{g1}}{N_{g1}} = \frac{\pi(50)}{25} = \frac{\pi(120)}{60} = 6.283 \text{ mm}, \quad p_d = \frac{N_{p1}}{d_{p1}} = \frac{N_{g1}}{d_{g1}} = \frac{25}{50} = \frac{60}{120} = 0.5 \text{ mm}^{-1},$$

$$m = \frac{d_{p1}}{N_{p1}} = \frac{d_{g1}}{N_{g1}} = \frac{50}{25} = \frac{120}{60} = 2 \text{ mm}.$$

(10.1)

For the second gear pair, we have

$$p_c = \frac{\pi d_{p2}}{N_{p2}} = \frac{\pi d_{g2}}{N_{g2}} = \frac{\pi(60)}{24} = \frac{\pi(125)}{50} = 7.854 \text{ mm}, \quad p_d = \frac{N_{p2}}{d_{p2}} = \frac{N_{g2}}{d_{g2}} = \frac{24}{60} = \frac{50}{125} = 0.4 \text{ mm}^{-1},$$

$$m = \frac{d_{p1}}{N_{p1}} = \frac{d_{g1}}{N_{g1}} = \frac{60}{24} = \frac{125}{50} = 2.5 \text{ mm}.$$

(10.2)

Therefore, the velocity ratio Z_r of the gear train is:

$$Z_r = \frac{\omega_{out}}{\omega_{in}} = \frac{N_{p1}}{N_{g1}} \frac{N_{p2}}{N_{g2}} = \frac{25}{60} \frac{12}{50} = \frac{1}{5}, \text{ or}$$

(10.3a)

$$Z_r = \frac{\omega_{out}}{\omega_{in}} = \frac{d_{p1}}{d_{g1}} \frac{d_{p2}}{d_{g2}} = \frac{50}{120} \frac{60}{125} = \frac{1}{5} \tag{10.3b}$$

where ω_{out} and ω_{in} are the output and input angular velocities of the gear train system, respectively; d_{p1}, d_{g1}, d_{p2}, and d_{g2} are the pitch circle diameters of the respective four gears; and N_{p1}, N_{g1}, N_{p2}, and N_{g2} are the number of teeth of the four respective gears. The velocity ratio of the gear train shown in Figure 1 is *1:5*, i.e., the angular velocity is reduced *5* times at the output shaft. Theoretically, the torque output will increase *5* times if there is no loss due to, e.g., friction. Note that we will use *MMGS* units system for this lesson.

SOLIDWORKS Part and Assembly Files

In this lesson, *SOLIDWORKS* parts and assemblies of the gear train have been created. There are six model files: *gbox_housing.SLDPRT*, *gbox_input.SLDPRT*, *gbox_middle.SLDPRT*, *gbox_output.SLDPRT*, *Lesson10.SLDASM*, and *Lesson10withResults.SLDASM*. We will start with *Lesson10.SLDASM*, in which the gears are fully assembled to the housing. In addition, the assembly file *Lesson10withResults.SLDASM* consists of a complete simulation model with simulation results. You may want to open *Lesson10withResults.SLDASM* to preview the gear motion.

Note that the housing part in *SOLIDWORKS* was converted directly from *PTC Creo* part. The three gear parts in *SOLIDWORKS* were converted from respective *PTC Creo* assemblies. There were nine, nine, and six distinct parts within the three gear assemblies, respectively. These three *PTC Creo* assemblies (and associated parts) were first converted to *SOLIDWORKS* as assemblies. Parts in each assembly were then merged into a single gear part in *SOLIDWORKS*. The detailed part and assembly conversions as well as merging multiple parts into a single part in *SOLIDWORKS* can be found in Appendix C.

In the *SOLIDWORKS* assembly model *Lesson10.SLDASM* (and *Lesson10withResults.SLDASM*), there are nine assembly mates. The first three mates, *Concentric1*, *Coincident1*, and *Coincident2*, assemble the input gear to the housing. The input gear is fully constrained. Note that before entering *Motion*, we will suppress *Coincident2* in order to allow the input gear to rotate along the Z-axis (see Figures 10-3a, b, and c). Similarly, the next three mates, *Concentric2*, *Coincident3* and *Coincident4*, assemble the middle gear to the housing. Again, we will suppress *Coincident4* to allow the middle gear to rotate about the Z-axis (see Figures 10-3d, e and f). The third set of mates, *Concentric3*, *Coincident5*, and *Coincident6*, does the same for the output gear (see Figures 10-3g, h and i). Similarly, *Coincident6* will be suppressed to allow for rotation. Note that the three suppressed mates are created to properly orient the three gears so that the gear teeth mesh properly between pairs. The three mates to be suppressed also define the initial configuration of the gear train.

As mentioned earlier, one important factor for the animation of gear motion to "look right" is to mesh the gear teeth properly. You may want to use the *Front* view and zoom in to the tooth mesh areas to check if the two pairs of gears mesh properly (see Figure 10-4). They should mesh well, which is accomplished by the three coincident mates that will be suppressed before entering *Motion*. Note that the tooth profile is represented by straight lines, instead of more popular ones such as involutes, for simplicity.

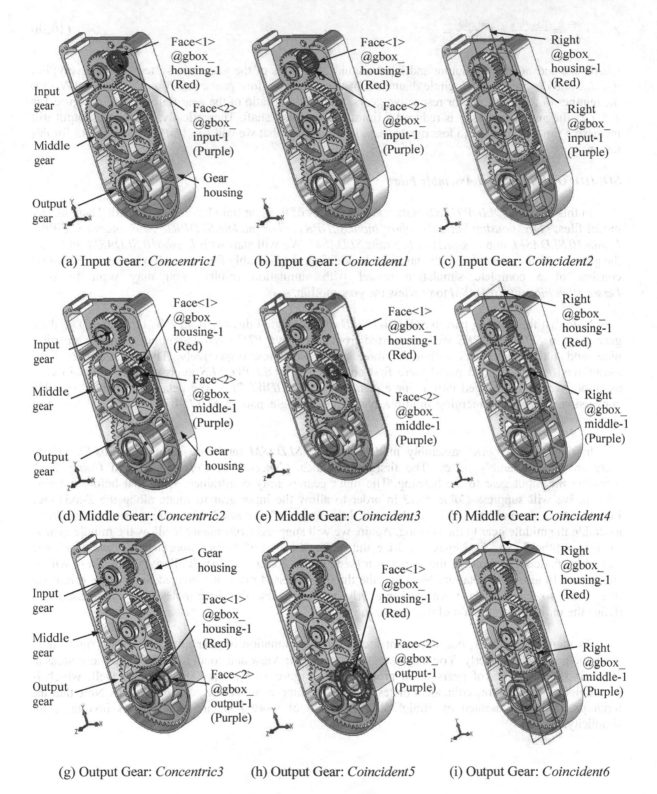

(a) Input Gear: *Concentric1*   (b) Input Gear: *Coincident1*   (c) Input Gear: *Coincident2*

(d) Middle Gear: *Concentric2*   (e) Middle Gear: *Coincident3*   (f) Middle Gear: *Coincident4*

(g) Output Gear: *Concentric3*   (h) Output Gear: *Coincident5*   (i) Output Gear: *Coincident6*

Figure 10-3 Assembly Mates Defined in *Lesson10.SLDASM*

If you turn on the axis display (*View* → *Hide/Show* → *Axes*), axes parallel to the Z-axis that pass through the center of the respective gears will appear (please see Figure 10-5). These axes are necessary for creating the gear mates in this lesson.

Gear housing

Input gear

Middle gear

Output gear

Figure 10-4 Gear Teeth Properly Meshed

Simulation Model

The gear housing is assigned as the ground part. All three gear subassemblies will rotate with respect to their respective axes. The four gears will be meshed into two gear pairs: *Pinion 1* with *Gear 1*, and *Pinion 2* with *Gear 2*, as discussed earlier. In *SOLIDWORKS*, gear pairs can be created by selecting two axes of the respective gears (or cylinders) using *Mechanical Mates* option. Note again that before the gear mates can be created, we will suppress the three coincident mates that orient the gears, i.e., *Coincident2*, *Coincident4* and *Coincident6*, in order to allow desired gear rotation motion. The axis of the input gear will be driven by a rotary motor with a constant angular velocity of *360* degrees/sec. We will conduct a kinematic analysis for this example.

Rotary Motor

Rotating axes

Figure 10-5 Gear Train Motion Model

10.3 Using *SOLIDWORKS Motion*

Start *SOLIDWORKS* and open *Lesson10.SLDASM*.

Before entering *Motion*, there are two things we need to do. First, we will suppress three assembly mates, *Coincident2*, *Coincident4* and *Coincident6*. Second, we will create two gear mates for the two gear pairs, respectively.

From *SOLIDWORKS* browser, expand the *Mates* node, click *Coincident2*, and choose *Suppress*. The mate *Coincident2* will become inactive. Repeat the same to suppress *Coincident4* and *Coincident6*. Save the model.

Next, turn on the axis view by choosing from the pull-down menu *View → Hide / Show → Axes*. All three axes, one for each gear assembly, will appear in the graphics area.

We will create two gear mates. Choose from the pull-down menu *Insert → Mate*. In the *Mate* dialog box, the *Mate Selections* field will be active, and is ready for you to pick entities.

Pick the axes of the input and middle gears from the graphics area. Choose *Mechanical Mates*, click *Gear*, and enter *50mm* and *120mm* for *Ratio*, as shown in Figure 10-6. Click the checkmark on top to accept the mate definition.

Note that if the axes of the two gears are pointing in the opposite direction, you will have to click *Reverse* box (right below the *Ratio* text field in the *Mate* dialog box) to correct the rotation direction. In this example, all three axes are pointing in the same direction. Therefore, do not choose *Reverse*.

Repeat the same steps to define the second gear mate. This time we pick the axes of the middle and output gears, and enter *60mm* and *125mm* for *Ratio*. Click the checkmark to accept the mate, and click it one more time to close the dialog box. Two new mates, *GearMate1(gbox_input<1>, gbox_middle<1>)* and *GearMate2(gbox_middle<1>,gbox_output<1 >)*, are now listed under *Mates* node.

Figure 10-6 Defining Gear Mate

Now we are ready to enter *Motion*.

Click the *Motion Study* tab at the bottom of the graphics area to bring up the *MotionManager* window. We will add a rotary motor and create a one-second simulation. You may want to expand the *Mates* node to make sure that *Coincident2*, *Coincident4*, and *Coincident6* stay suppressed. Sometimes, *SOLIDWORKS* unsuppresses mates while entering *Motion*. If this happens, we need to suppress the three mates.

Adding a Rotary Motor

Click the *Motor* button 🐾 from the *Motion* toolbar to bring up the *Motor* dialog box (Figure 10-7). Choose *Rotary Motor* (default). Move the pointer to the graphics area, and pick *Axis1* of the input gear, as shown in Figure 10-8. A circular arrow appears indicating the rotational direction of the rotary motor. A counterclockwise direction (along Z-axis) is desired. You may change the direction by clicking the direction button ⬈ right below *Component/Direction,* if necessary. Choose *Constant speed* and enter *60* RPM for speed. Click the checkmark ✔ on top of the dialog box to accept the motor definition. You should see a *RotaryMotor1* added to the *MotionManager* tree.

Figure 10-7   Figure 10-8

We are ready to run a simulation. If you are just interested in gear motion animation, you may simply choose *Animation* option for the study. However, if you want to see motion plots, you will need to choose *Motion Analysis* for study.

Running Simulation

Choose *Motion Study* for the study, and drag the end time key to one-second mark. Click the *Motion Study Properties* button ⚙ from the *Motion* toolbar. In the *Motion Study Properties* dialog box, enter *60* for *Frames per second*, and click the checkmark on top. Click the *Calculate* button 🖩 from the *Motion* toolbar to simulate the motion. A one-second simulation will be carried out. After a short moment, you should see the gears start moving. The input gear rotates *360* degrees as expected, which drives all other gears to rotate.

You may want to adjust the playback speed to 10% (0.1x) to slow down the animation since the total simulation duration is only one second. Look closer to the gear teeth. Both gear pairs should mesh correctly. The animation looks good. All gear teeth mesh properly.

Saving and Reviewing Results

We will graph the angular velocity of the output gear. In order to show result graphs, we will have to choose *Motion Analysis* for the study, and rerun the simulation. Click the *Motion Study Properties* button ⚙ , enter *60* for *Frames per second*, and click the checkmark on top. Click the *Calculate* button 🖩 from the *Motion* toolbar to simulate the motion.

Right click *gbox_output<1>* in the *MotionManager* tree, and choose *Create Motion Plot*. In the *Results* dialog box, choose *Displacement/Velocity/Acceleration*, select *Angular Velocity*, and then *Z Component*. The graph will appear similar to that of Figure 10-9, which shows that the output velocity is a constant of *72* degrees/sec. Note that this magnitude is one fifth of the input velocity since the velocity ratio is *1:5*. Both the input (*Pinion 1*) and output gears (*Gear 2*) rotate in the same direction. *Motion* gives good results.

Save your model for future reference.

Figure 10-9 Angular Velocity of Output Gear

Exercises:

1. The same gear train will be used for this exercise. Create a constant torque for the input gear (*gbox_input.SLDPRT*) about the *Z*-axis. Turn on friction for all three axles (*Steel-Dry/Steel-Dry*). Define and run a *2*-second dynamic simulation for the gear train.

    (i)   What is the minimum torque that is required to rotate the input gear, and therefore, the entire gear train?

    (ii) If the torque applied to the input gear is *100* mm-N, what is the output angular velocity of the gear train at the end of the *2*-second simulation? Verify the simulation result using your own calculation.

    (iii) Create a graph for the reaction moment between gears of the first gear pair (*GearMate1*) due to the *100* mm-N torque. What is the reaction moment obtained from simulation?

Notes:

Lesson 11: Planetary Gear Train Systems

11.1 Overview of the Lesson

In this lesson we discuss planetary gear train systems. Planetary gearing is a gear system that is composed of one or more outer gears, or *planet* gears, rotating about a central, or *sun* gear, as shown in Figure 11-1. Typically, the planet gears are mounted on a movable arm (or carrier) which may rotate relative to the sun gear. A planetary gear system may also incorporate the use of an outer *ring* gear (or annulus), which meshes with the planet gears. In general, one of the most important tasks in designing planetary gear train systems is to achieve a required gear reduction in a compact space.

Planet Gear 1 —
Ring Gear (or Annulus)

Arm
(or Carrier)
Sun Gear

Bearing

Planet Gear 2

Planet Gear 3

Figure 11-1 Planetary Gear Train

In many planetary gearing systems, one of four basic components (that is, sun gear, planet gears, arm, and ring gear) is held stationary (usually the ring gear is fixed); one of the remaining components is an *input*, receiving power to the gear train system, while the last component is an *output*, disseminating power from the system. The ratio of input rotation to output rotation is dependent upon the number of teeth in meshed gear pairs, and upon which component is held stationary. One situation is when the ring gear is held stationary, and the sun gear is used as input. In this case, the planet gears rotate around the sun gear at a rate determined by the number of teeth in meshed gear pairs, i.e., sun/planet and planet/ring gears.

Planetary gear trains are commonly employed in mechanical systems. They are primarily designed for large gear reductions, hence building up sufficient torque to drive mechanism system. One of the common places to find the planetary gear trains is in the power screwdriver (Figure 11-2), where a large gear reduction is necessary for a large torque output in a small space. Series of planetary gear trains are often packaged to receive a maximum gear reduction.

In this lesson, we will learn how to simulate motion of planetary gear trains and calculate the velocity ratios. We will discuss the theory and verify the simulation results obtained from *Motion*. It is important that you understand the theory and are able to calculate the velocity ratio of the gear train by yourself.

11.2 The Planetary Gear Train Examples

Physical Model

We discuss two examples. The first example is a single planetary gear train, as shown in Figure 11-3. This gear train consists of only one planet gear mounted on the arm. The planet gear is meshed with a stationary ring gear. In this system, there is only one pair of meshed gears: the planet and the ring gears. The pitch circle diameters of the planet and ring gears are $d_p = 2.333$ and $d_r = 8.167$ in., respectively. There are 12 and 42 teeth on the planet and ring gears, respectively. Therefore, the circular pitch P_c and diametral pitch P_d of the gears are, respectively,

Figure 11-2 Planetary Gear Trains
Found in Power Screw Driver

$$p_c = \frac{\pi d_p}{N_p} = \frac{\pi d_r}{N_r} = \frac{\pi(2.333)}{12} = \frac{\pi(8.167)}{42} = 0.6108 \text{ in.} \tag{11.1a}$$

$$p_d = \frac{N_p}{d_p} = \frac{N_r}{d_r} = \frac{12}{2.333} = \frac{42}{8.167} = 5.143 \text{ in.}^{-1} \tag{11.1b}$$

Figure 11-3 The Single Planetary
Gear Train System

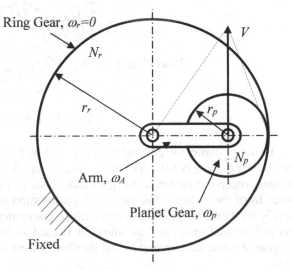

Figure 11-4 Velocity Analysis for the Single
Planetary Gear Train System

Note that every single counterclockwise turn of the arm produces $(1-N_r/N_p)$ clockwise turns of the planet gear. This can be figured out from velocity analysis. For example, as shown in Figure 11-4, the linear velocity at the end of the arm where it connects to the center of the planet gear is $V = (r_r-r_p)\omega_A$. Since the planet gear is "rolling" along the ring gear, the linear velocity at its center, which is the same as V, is $V = r_p\omega_p$. Therefore,

$$V = (r_r - r_p)\omega_A = r_p(-\omega_p) \tag{11.2a}$$

$$\omega_p = \left(1 - \frac{r_r}{r_p}\right)\omega_A = \left(1 - \frac{N_r}{N_p}\right)\omega_A \tag{11.2b}$$

where the fact that the arm and the planet gear rotate in the opposite direction has been considered in Eq. 11.2a with a negative sign placed in front of the angular velocity of the planet gear ω_p. The velocity ratio Z_r of the gear train is:

$$Z_r = \frac{\omega_{out}}{\omega_{in}} = \frac{\omega_p}{\omega_A} = 1 - \frac{N_r}{N_p} = 1 - \frac{42}{12} = -2.5 \tag{11.3}$$

That is, one counterclockwise turn of the arm produces 2.5 clockwise turns on the planet gear. If the arm is rotating at 360 degrees/sec counterclockwise, the planet gear will rotate at –900 (that is, –2.5×360) degrees/sec (clockwise).

The second example, as shown in Figure 11.5, consists of an additional sun gear and two additional planet gears. Again, the ring gear is stationary and in this case the sun gear is driven by a rotary motor of constant speed. The pitch circle diameter of the sun gear is $d_s = 3.5$ in. and the number of teeth is $N_s = 18$. Therefore, the circular pitch P_c and diametral pitch P_d of the sun gear are 0.6108 in. and 5.143 in.$^{-1}$, respectively, which are identical to the remaining gears in the system, as expected.

Note that for a general planetary gear train system, the arm is usually not directly driven by the power source (such as a motor). The general equation for calculating the velocity ratio is the following:

Figure 11-5 The Multiple Planetary Gear Train System

$$Z_r = \frac{\omega_{out} - \omega_A}{\omega_{in} - \omega_A} = \Pi\left(\pm\frac{N_{driving}}{N_{driven}}\right) \tag{11.4}$$

where $N_{driving}$ and N_{driven} are the number of teeth on the driving and driven gears, respectively, of a gear pair; and Π is an operator that creates a product for the terms enclosed in the parentheses.

Note that the driving and driven gears can be identified by power input and output within each gear pair. For example, for the gear system shown in Figure 11-5, if the sun gear is directly connected to a motor, then the sun gear will be the driving gear and the planet gear (any one of the three) is the driven gear. In the second gear pair, i.e., the planet and the ring gears, the planet gear will be the driving gear and the ring gear will be the driven gear. The sign on the right-hand side of Eq. 11.4 will be determined by how the gear pairs are meshed, externally or internally. For a gear pair consisting of two regular spur gears, such as the one discussed in *Lesson 10*, the sign is negative since the two spur gears are meshed externally and rotate in the opposite directions. For a gear pair including a ring gear, the sign will be positive since, in this case, the planet and the ring gears are meshed internally and rotate in the same direction. Therefore, for the second example, if the sun gear is driven by a motor with a constant angular velocity of *360* degrees/sec counterclockwise, what will be the velocity ratio of the system? Well, we know the output gear, i.e., the ring gear, is stationary; therefore, $\omega_{out} = 0$. In order to calculate the velocity ratio, we will have to figure out the angular velocity of the arm first. This can be accomplished by using Eq. 11.4.

$$\frac{\omega_{out} - \omega_A}{\omega_{in} - \omega_A} = \frac{0 - \omega_A}{360 - \omega_A} = \left(-\frac{N_s}{N_p}\right)\left(\frac{N_p}{N_r}\right) = \left(-\frac{18}{12}\right)\left(\frac{12}{42}\right) = -0.4286 \tag{11.5}$$

Therefore, $\omega_A = 108.0$ degree/sec counterclockwise, and the velocity ratio is *–0.4286* if the input and output gears of the system are the sun and ring gears, respectively.

If the sun gear is the input gear and the planet gear is the output gear, then the angular velocity of the planet gear can be calculated by using Eq. 11.4, i.e.,

$$\frac{\omega_{out} - \omega_A}{\omega_{in} - \omega_A} = \frac{\omega_p - 108}{360 - 108} = \left(-\frac{N_s}{N_p}\right) = \left(-\frac{18}{12}\right) = -1.5 \tag{11.6}$$

Therefore, $\omega_p = -270.0$ degrees/sec (clockwise), and the velocity ratio is *–1.5*.

SOLIDWORKS Part and Assembly Files

In this lesson, *SOLIDWORKS* parts and assemblies of the gear train have been created. There are eight model files: *bearing.SLDPRT*, *arm.SLDPRT*, *planet.SLDPRT*, *ring.SLDPRT*, *Lesson11SingleGearTrain.SLDASM*, *Lesson11SingleGearTrainwithRresults.SLDASM*, *Lesson11MultiGearTrain.SLDASM*, and *Lesson11MultiGearTrainwithResults.SLDASM*. We will start with *Lesson11SingleGearTrain.SLDASM* (see Figure 11-3) to carry out motion simulation for a single planetary gear train system. We will then use *Lesson11MultiGearTrain.SLDASM* (see Figure 11-5) to do the same for a multi-planetary gear train system. The assembly files *Lesson11SingleGearTrainwithResults.SLDASM* and *Lesson11MultiGearTrainwithResults.SLDASM* consist of complete simulation models with simulation results for the single and multi-planetary gear train systems, respectively. You may want to open these two assembly files to preview the motion of the respective two gear train systems.

Gear Parts

Note that the planet gear, sun gear, and ring gear of the *SOLIDWORKS* models were imported directly from respective *PTC Creo* parts. Pitch circles of diameters *2.333*, *3.500*, and *8.167*, for the planet, sun, and ring gears are added as sketches to the respective *SOLIDWORKS* parts, as illustrated in Figure 11-6. Please note that axes are not selected for defining gear pairs (like that of *Lesson 10*); instead, pitch circles created in sketches are selected to define gear mates in *Motion*. Incorrect simulation results were found if axes are selected for defining the gear pairs in this example.

Single Planetary Gear Train Assembly

In the *SOLIDWORKS* assembly models *Lesson11SinglePlaneteryGearTrain.SLDASM* (and *Lesson11SinglePlaneteryGearTrainwithResults.SLDASM*), there are nine assembly mates. When you open the assembly, you should see the assembly mates listed in the *FeatureManager* tree (or *SOLIDWORKS* browser) like that of Figure 11-7.

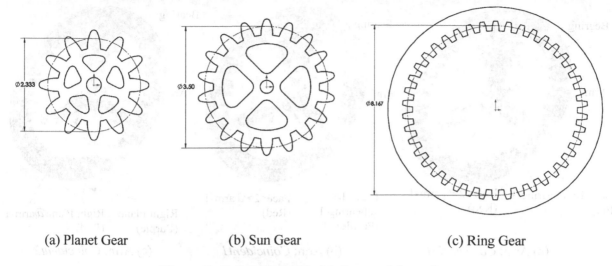

(a) Planet Gear   (b) Sun Gear   (c) Ring Gear

Figure 11-6 Gear Solid Models Imported from *Creo*

The first three mates, *Concentric1*, *Coincident1*, and *Coincident2,* assemble the arm to the bearing (see Figures 11-8a, b and c), which is fixed. The arm is fully constrained. In *Motion* analysis, we will suppress *Coincident2* to allow the arm to rotate about the bearing in *Z*-axis.

Similarly, the next three mates, *Concentric2*, *Coincident3*, and *Coincident4*, assemble the planet gear to the arm (see Figures 11-8d, e and f). Again, we will suppress *Coincident4* to allow the planet gear to rotate in the *Z*-direction.

The third set of mates, *Concentric3*, *Distance1*, and *Coincident5*, does the same for the ring gear (see Figures 11-8g, h and i).

In *Motion* analysis, ring gear will be fully constrained. Note that the two mates, *Coincident2* and *Coincident4,* to be suppressed in motion analysis are created to orient the gears so that the gear teeth mesh properly between pairs (see Figure 11-9). The two mates to be suppressed also define the initial configuration of the gear train.

Figure 11-7 Assembly Mates of *Lesson11SinglePlaneteryGearTrain.SLDASM* listed in the *FeatureManager* tree

Multi-Planetary Gear Train Assembly

In the *SOLIDWORKS* assembly model *Lesson11MultiGearTrain.SLDASM* (and *Lesson11MultiGearTrainwithResults.SLDASM*), there are eighteen assembly mates. When you open the assembly, you should see the assembly mates listed in the *FeatureManager* tree like that of Figure 11-10.

Bearing

Arm

Face<1>@bearing-1   Face<2>@arm-1
(Purple)   (Red)

(a) Arm: *Concentric1*

Bearing

Arm

Face<1>   Face<2>@arm-1
@bearing-1   (Red)
(Purple)

(b) Arm: *Coincident1*

Bearing

Arm

Right Plane   Right Plane@arm-1
(Purple)   (Red)

(c) Arm: *Coincident2*

Planet
Gear

Arm

Face<1>@arm-1, and
Face<2>@planet-1

(d) Planet Gear: *Concentric2*

Planet
Gear

Arm

Face<2>   Face<1>@arm-1
@planet-1   (Purple)
(Red)

(e) Planet Gear: *Coincident3*

Planet
Gear

Arm

Right Plane   Right Plane
@arm-1   @planet-1 (Red)
(Purple)

(f) Planet Gear: *Coincident4*

Bearing

Ring
Gear

Face<1>   Face<2>@bearing-1
@ring-1 (Red)   (Purple)

(g) Ring Gear: *Concentric3*

Planet
Gear

Arm

Face<2>   Face<1>@arm-1
@planet-1   (Purple)
(Red)

(h) Ring Gear: *Distance1*

Bearing

Ring
Gear

Top Plane   Top Plane@bearing-1
@ring-1   (Purple)
(Red)

(i) Ring Gear: *Coincident5*

Figure 11-8 Assembly Mates Defined in *Lesson11SingleGearTrain.SLDASM*

Figure 11-9 Teeth of Planet and Ring Gears Properly Meshed

The first nine mates are identical to those of *Lesson11SingleGearTrain.SLDASM*, with which the arm, planet gear, and ring gear are fully assembled. As mentioned above, we suppress *Coincident2* and *Coincident4* to allow proper motion.

The next three mates, *Concentric4*, *Coincident6*, and *Angle1* assemble the second planet gear to the assembly (see Figure 11-11a, b, and c), similar to that of the first planet gear. Note that *Angle1* mate orients the second planet gear so that its gear teeth mesh well with the ring gear. In *Motion* analysis, we will suppress *Angle1* to allow the planet to rotate in respect to the arm about the *Z*-axis.

The third planet gear is assembled to the assembly using three mates, *Concentric5*, *Coincident7*, and *Angle2*, similar to those of the second planet gear.

The sun gear is assembled to the assembly by defining three assembly mates (see Figure 11-11d, e, and f) following similar approach as that of the planet gears. The three mates are *Concentric6*, *Coincident8*, and *Angle3*. Again, *Angle3* mate orients the sun gear so that its gear teeth mesh well with the planet gears. In *Motion* analysis, we will suppress *Angle3* to allow the sun gear to rotate in respect to the arm about the *Z*-axis.

Figure 11-10 Assembly Mates of *Lesson11MultiPlaneteryGearTrain.SLDASM* listed in the *FeatureManager* tree

Note that in *Motion* analysis, ring gear will be fully constrained. Five mates, *Coincident2*, *Coincident4*, *Angle1*, *Angle2*, and *Angle3* that are created for an initial condition where all gear teeth are properly meshed (see Figure 11-12) are suppressed. In the *FeatureManager* tree, these suppressed mates are grayed out, as seen in Figure 11-10.

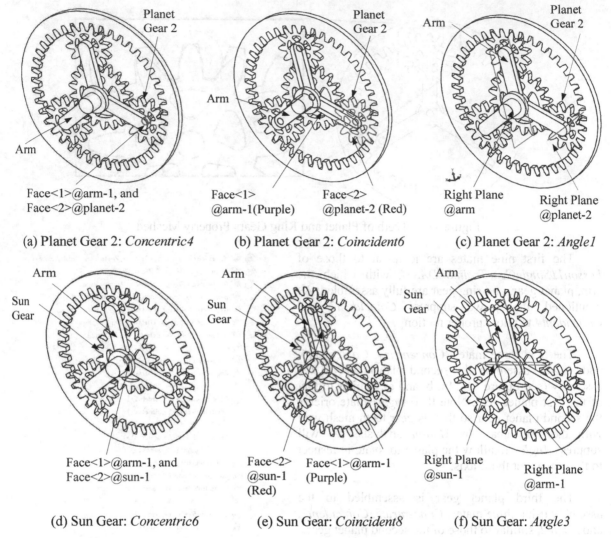

(a) Planet Gear 2: *Concentric4*   (b) Planet Gear 2: *Coincident6*   (c) Planet Gear 2: *Angle1*

(d) Sun Gear: *Concentric6*   (e) Sun Gear: *Coincident8*   (f) Sun Gear: *Angle3*

Figure 11-11 Assembly Mates Defined in *Lesson11MultiPlaneteryGearTrain.SLDASM*

Figure 11-12 Teeth of Sun, Ring, and Planet Gears Properly Meshed

Simulation Models

Two motion simulation models are to be created for the single and multi-planetary gear train systems, respectively.

For the single planetary gear train system, the bearing and ring gear are fixed (defined as the ground body). A gear mate is defined for the planet gear and the ring gear by selecting their respective pitch circles (see Figure 11-13a). The arm is rotating CCW driven by a rotary motor of *60* RPM constant angular velocity (hence, $\omega_A = -360°$/sec along the Z-direction). Note again that we will suppress the two coincident mates that orient the gears, i.e., *Coincident2*, and *Coincident4*, in order to allow for desired gear rotation motion.

For the multi-planetary gear train system (see Figure 11-13b), the bearing and the ring gear are fixed (ground body) like that of the single gear train system. Two more planet gears and one sun gear are added to the system. Three identical gear mates are defined for the respective three planet gears and the ring gear by selecting their respective pitch circles. Another gear mate is defined for the first planet gear and the sun gear by selecting their respective pitch circles. Note that adding gear mates between the remaining planet gears and the sun gear will over-constrain the gear train system. Instead of rotating the arm, the sun gear is rotating CCW driven by a rotary motor of *60* RPM constant angular velocity (hence, $\omega_S = -360°$/sec along the Z-direction). Note again that we suppress five coincident mates that orient the gears, i.e., *Coincident2*, *Coincident4*, *Angle1*, *Angle2*, and *Angle3*, in order to allow desired gear rotation motion. Note that in this lesson, we will use the default unit system, i.e., *IPS*.

11.3 Using *SOLIDWORKS Motion*: Single Planetary Gear Train

Start *SOLIDWORKS* and open *Lesson11SingleGearTrain.SLDASM*.

Before entering *Motion*, we first suppress two assembly mates, *Coincident2* and *Coincident4*. From the *FeatureManager* tree, expand the *Mates* node, click *Coincident2*, and choose *Suppress*. The mate *Coincident2* will become suppressed (turned into gray color). Repeat the same to suppress *Coincident4*. Save the model under a different name.

    (a) Single Planetary Gear Train System   (b) Multi- Planetary Gear Train System

Figure 11-13 Gear Train Motion Simulation Models

Next, show sketches by choosing from the pull-down menu *View → Hide / Show → Sketches*. The two sketches of the respective gear pitch circles appear in the graphics area.

We check if the *IPS* is chosen by choosing from the pull-down menu *Tools → Options*. In the *Document Properties* dialog box (Figure 11-14), select the *Documents Properties* tab and select *Units*. Select *IPS (inch, pound, second)*. We will stay with this unit system for this lesson.

Also, we increase the decimals from the default 2 to 3 digits since the pitch circle diameters are down to tens of thousandth of an inch. To increase the decimals to 3 digits, you may click the dropdown button to the right of the cell in the *Length* row under the *Decimals* column (circled in Figure 11-14). Pull down the selection and choose *.123* for 3 digits. Click *OK* to accept the change.

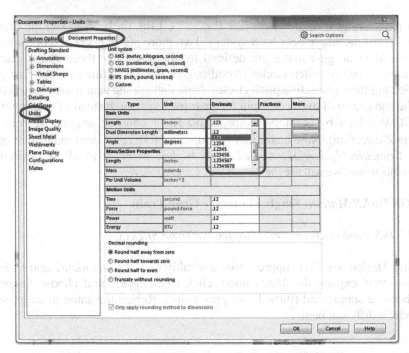

Figure 11-14 The *Document Properties* Dialog Box

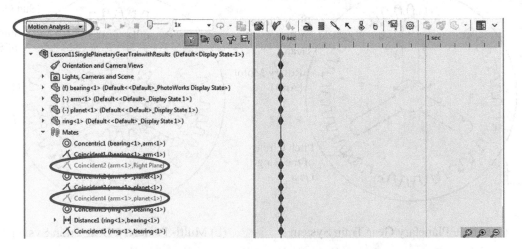

Figure 11-15 The *MotionManager* Window

Click the *Motion Study* tab at the bottom of the graphics area to bring up the *MotionManager* window. You should see that *Motion Analysis* is selected (circled in Figure 11-15), and the two mates stay in gray color (suppressed, circled in Figure 11-15).

Adding a Gear Mate

We create a gear mate next. Choose from the pull-down menu *Insert → Mate*. In the *Mate* dialog box (see Figure 11-16a), the *Mate Selections* field will be active, and is ready for you to pick entities. Select the pitch circles of the planet and ring gears from the graphics area (see Figure 11-16b). Choose *Mechanical Mates*, click *Gear*, the pitch circle dimeters of the planet and ring gears, *2.333* in. and *8.167* in. appear under *Ratio*, as shown in Figure 11-16a. The *Reverse* box should be checked automatically. Note that the planet and ring gears rotate in the same direction; the *Reverse* box should have been checked off. However, if the box is checked off, in simulation, the gears rotate in the opposite direction. Therefore, the box must be checked.

Click the checkmark ✅ on top of the dialog box to accept the mate definition. A local gear mate is now added to *MotionManager* tree listed under *Mates* (see Figure 11-17).

Adding a Rotary Motor

Click the *Motor* button 🦀 from the *Motion* toolbar to bring up the *Motor* dialog box (Figure 11-18a). Choose *Rotary Motor* (default). Move the pointer to the graphics area, and pick a circular arc of the arm, as shown in Figure 11-18b.

(a) *Gear Mate* Dialog Box

(b) Pick Pitch Circles from Graphics Area

Figure 11-16 Defining Gear Mate

A circular arrow appears indicating the rotational direction of the rotary motor. A counterclockwise direction (along Z-axis) is desired. Change the direction by clicking the direction button ⬈ right below *Component/Direction* if needed. Choose *Constant speed* and enter *60* RPM for speed. Click the checkmark ✔ on top of the dialog box to accept the motor definition. You should see a *RotaryMotor1* added to the *MotionManager* tree. We are ready to run a simulation.

Running Simulation

Choose *Motion Analysis* for the study, and drag the end time key to the one-second mark. Click the *Motion Study Properties* button ⚙ from the *Motion* toolbar. In the *Motion Study Properties* dialog box, enter *100* for *Frames per second*, and click the checkmark on top. Click the *Calculate* button 🖩 from the *Motion* toolbar to simulate the motion. A one-second simulation will be carried out. After a short moment, you should see the arm and the planet gear start moving. The arm rotates *360* degrees as expected, which drives the planet gear to rotate.

You may want to adjust the playback speed to 10% (*0.1x*) to slow down the animation since the total simulation duration is only one second. Look closer at the gear teeth. Gear teeth of the planet and the ring gears should mesh correctly throughout the simulation. The simulation looks good.

Figure 11-17 A Local Gear Mate Added to the *MotionManager* tree

(a) The *Motor* Dialog Box

(b) Pick a Circular Arc from Graphics Area

Figure 11-18 Defining a Rotary Motor

Reviewing Results

We will graph the angular velocity of the output gear.

Right click *planet<1>* in the *MotionManager* tree, and choose *Create Motion Plot*. In the *Results* dialog box, choose *Displacement/Velocity/Acceleration*, select *Angular Velocity*, and then *Z Component*. Click the checkmark to accept the motion plot.

A graph will appear similar to that of Figure 11-19, which shows that the output velocity is a constant of *900* degrees/sec. Note that this magnitude is *2.5* times of the input velocity at the arm since the velocity ratio is *–2.5*. The input (*arm*) and output gear (*planet*) rotate in the opposite direction. *Motion* gives good results.

Figure 11-19 Angular Velocity of Planet Gear

You may save the model, close it, and move on to the next motion example.

11.4 Using *SOLIDWORKS Motion*: Multi-Planetary Gear Train

Open *Lesson11MultiPlanetaryGearTrain.SLDASM*.

Before entering *Motion*, we make sure the five assembly mates are suppressed, *Coincident2*, *Coincident4*, *Angle1*, *Angle2*, and *Angle3*.

Similar to the previous case, we display sketches, and increase the decimals from the default 2 to 3 digits.

Click the *Motion Study* tab at the bottom of the graphics area to bring up the *MotionManager* window. You should see that *Motion Analysis* is selected (circled in Figure 11-20), and the five mates stay in gray color (suppressed, circled in Figure 11-20).

Adding Gear Mates for Planet and Ring Gears

We will first create three gear mates for the respective three planet gears and the ring gear.

Choose from the pull-down menu *Insert → Mate*. In the *Mate* dialog box (like that of Figure 11-16), pick the respective pitch circles, for example, planet gear 1 and ring gear, like before. Choose *Mechanical Mates*, click *Gear*, the pitch circle diameters of the planet and ring gears, *2.333* in. and *8.167* in., appear under *Ratio*. The *Reverse* box should be checked automatically.

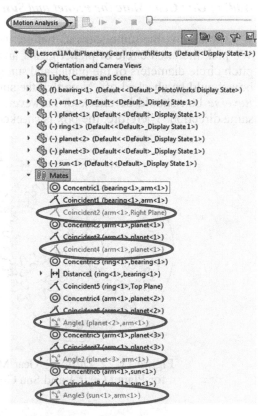

Figure 11-20 The Assembly Mates Listed in the *MotionManager* tree

Again, we keep the *Reverse* box checked. Click the checkmark ✔ on top of the dialog box to accept the mate definition.

Repeat the same for planet gear 2 and 3 (see Figure 11-21). Three local gear mates are now added to the *MotionManager* tree listed under *Mates*.

    (a) Planet Gear 1   (b) Planet Gear 2   (c) Planet Gear 3

Figure 11-21 Defining Gear Mates for Planet Gears and Ring Gear

Adding One Gear Mate for Planet and Sun Gears

Following the same steps as before, we add only one gear mate for planet gear 1 and the sun gear. We pick the pitch circle of planet gear 1 and pick the pitch circle of the sun gear (see Figure 11-22). The pitch circle diameters of the planet and sun gears, *2.333* in. and *3.50* in., appear under *Ratio*. The *Reverse* box should be checked off. Note that the sun and planet gears rotate in opposite directions; in theory, the *Reverse* box should be checked. However, if the box is checked, in simulation, the gears rotate in the same direction. Therefore, the box is checked off.

Figure 11-22 Defining Gear Mate
for Planet Gear 1 and Sun Gear

Figure 11-23 A Total of Four Gear Mates
Added to the *MotionManager* tree

The newly defined local gear mate is now added to the *MotionManager* tree listed under *Mates* (see Figure 11-23). We will not add more gear mates between the remaining planet gears and sun gear since additional gear mates will over-restrain the motion of the system.

Adding a Rotary Motor

We follow the same steps as before to create a rotary motor by picking the pitch circle of the sun gear from the graphics window (see Figure 11-24), entering *60* RPM, and reverse the direction for a CCW rotation. We are ready to run a simulation.

Running Simulation

Choose *Motion Analysis* for the study, and drag the end time key to the three-second mark. Click the *Motion Study Properties* button ⚙ from the *Motion* toolbar. In the *Motion Study Properties* dialog box, enter *100* for *Frames per second*, and click the checkmark on top. Click the *Calculate* button 🖩 from the *Motion* toolbar to simulate the motion. A three-second simulation will be carried out.

After a short moment, you should see the gears and arm start moving. The sun gear rotates *CCW* as expected, which drives all the other gears and therefore, the arm, to rotate. The simulation looks good. All gear teeth are meshed properly throughout the simulation.

Pick this pitch circle (Sun Gear)

Figure 11-24 Pick Pitch Circle of the Sun Gear to Define the Rotary Motor

Reviewing Results

We graph the angular velocity of the planet gear 1.

Following the same steps before, we right click *planet<1>* in the *MotionManager* tree, and choose *Create Motion Plot*. In the *Results* dialog box, choose *Displacement/Velocity/Acceleration*, select *Angular Velocity*, and then *Z Component*.

The graph will appear similar to that of Figure 11-25, which shows that the output velocity is a constant of *270* degrees/sec. Note that this magnitude is 1.5 times smaller than that of the input velocity at the arm since the velocity ratio is *–1.5*. The input (*sun*) and output gears (*planet*) rotate in the opposite direction. *Motion* gives good results.

Figure 11-25 Angular Velocity of Planet Gear

We have completed this lesson; you may save the model for future reference.

Exercises:

1. Use the same multi-planetary gear train example we created for this exercise. Delete the rotary motor that drives the sun gear. Add a new rotary motor to drive the arm at *60* RPM CCW. Plot the angular velocity for both the planet gear and the sun gear. Are the angular velocities obtained from *Motion* analysis accurate? How do you verify the results?

Lesson 12: Cam and Follower

12.1 Overview of the Lesson

In this lesson, we learn the cam and follower (or cam-follower) connection, or the *CamMateTangent* mate in *SOLIDWORKS*. Physically, a cam-follower is a mechanism that converts rotary motion into a linear motion. In *Motion*, a cam and follower can be assembled using a cam-follower assembly mate (or *CamMateTangent*). A simple form of a cam is a rotating disc with a variable radius so that its profile is not circular but oval or egg-shaped. When the disc rotates, its edge (or side face) pushes against a follower, which may be a small cylinder at the end of a lever or the end face of the lever or rod itself. The follower will thus rise and fall at exactly the same amount as the variation in radius of the cam. By profiling a cam appropriately, a desired cyclic pattern of straight-line motion, in terms of position, velocity, and acceleration, can be produced.

We will learn to create a motion model and simulate the control of opening and closing of an inlet or exhaustive valve, usually found in an internal combustion engine, using cam-follower connections. In a design such as that of Figure 12-1, the drive for the camshaft is taken from the crankshaft through a timing chain, which keeps the cams synchronized with the movement of the piston so that the valves open and close at a precise instant. The mechanism we will be working with consists of bushings, camshaft, pushrod, rocker, valve, valve guide and spring, as shown in Figure 12-1. The cam-follower connects the camshaft and the pushrod. When the cam on the camshaft pushes the pushrod up, the rocker rotates and pushes the valve on the other side downward. The spring surrounding the valve gets compressed and opens up the inlet for air to flow into the combustion chamber (or exhaustive to let the air out).

Figure 12-1 The Mechanism of Engine Inlet or Outlet Valve

First bushing
Rocker
Spring
Valve guide
Valve
Pushrod
Second bushing
Motion driver
Camshaft

12.2 The Cam and Follower Example

Physical Model

The camshaft and the rocker will rotate about the axes of their respective revolute joints connecting them to their respective bearings (defined as ground body). The camshaft is driven by a motor of constant velocity of *600* rpm (or *10* rev/sec). The profile of the cam consists of two circular arcs of *0.25* and *0.5* in. radii, respectively, as shown in Figure 12-2. The lower arc is concentric with the shaft, and the center of the upper arc is *0.52* in. above the center of the shaft. When the camshaft rotates, the cam mounted on the shaft pushes the pushrod up by up to *0.27* in. (that is, *0.52+0.25–0.5 = 0.27*). As a result, the rocker will rotate and push the valve at the other end downward at a frequency of *10* times/sec. The valve will move up to *0.27* in. downward since the pushrod and the valve are positioned at an equal distance from the rotation axis of the rocker. When the camshaft rotates where the larger circular arc (*0.5* in. radius) of the cam is in contact with the follower (in this example, the pushrod), the pushrod has room to move downward. At this point, the rocker will rotate back since the spring is uncompressed. As a result, the valve will move up, and therefore, close the inlet. The valve will be open for about *120* degrees per cycle, based on the geometric shape of the cam shown in Figure 12-2.

The units system chosen for this example is *IPS* and all parts are made of steel.

SOLIDWORKS Part and Assembly Files

The cam-follower system consists of seven parts, bushing (two), valve guide, camshaft, pushrod, rocker, and valve, as shown in Figure 12-1. In addition, there are two assembly files, *Lesson12.SLDASM* and *Lesson12withResults.SLDASM*, that you may download from the publisher's web site.

Figure 12-2 The Cam Profile

We will start with *Lesson12.SLDASM*, in which the parts are adequately assembled. In this assembly the first bushing is anchored (ground) and the second bushing and the valve guide are fully constrained. These three parts will be fixed (ground). The remaining four parts will be movable.

Same as before, the assembly file *Lesson12withResults.SLDASM* contains a complete simulation model with simulation results. You may want to open the assembly to see the motion animation of the mechanism. In the assembly where a motion model is completely defined, a mate (*Parallel2*) has been suppressed to allow movement between components. You can also see how the parts move by right clicking one of the moving parts in the graphics area, choosing *Move Component*, and dragging the part; for example, the camshaft to rotate with respect to the second bushing. The whole mechanism should move accordingly. Make sure you undo the part dragging to bring the assembly to its initial configuration.

There are eighteen assembly mates, including four coincident, three concentric, seven distance, two parallel, one tangent, and one cam-mate-tangent, as listed in the browser (see Figure 12-3). You may want to expand the *Mates* node in *SOLIDWORKS* browser to review the assembly mates. Click any of the mates; you should see the entities selected get highlighted in the graphics area.

As mentioned earlier, the first bushing, *bushing<1>*, shown in the browser is fixed to the assembly. The second bushing (*bushing<2>*) and the valve guide (*valve guide<1>*) were fully assembled to *bushing<1>*. The first three mates, *Coincident1*, *Distance1*, and *Distance2*, were employed to assemble *bushing<2>* to *bushing<1>*. And the next three distance mates assemble the valve guide to *bushing<1>*.

Figure 12-3 Assembly Mates Listed in the Browser

Note that the distance mates are essentially coincident mates with distance between entities. The distance mates were created to properly position the second bushing with respect to the first bushing. All three parts will be regarded as the ground body in *Motion*.

The next two mates, *Concentric1* and *Coincident2*, assemble the rocker (*rocker<1>*) to the first bushing (*bushing<1>*), allowing a rotation degree of freedom about the Z-axis, as shown in Figures 12-4a and b. *Motion* will map a revolute joint between the rocker and the first bushing.

The next part assembled is the pushrod (*pushrod<1>*). The pushrod was assembled to the rocker using *Tangent1*, *Distance6*, and *Coincident3* mates, as shown in Figures 12-4c, d, and e, respectively. As a result, the pushrod is allowed to move vertically at a distance of *1.25* in. from the *Right* plane of the first bushing (*Distance6*), at the same time maintaining tangency between the top of the cylindrical surface of the pushrod and the socket surface of the rocker.

The next part is the camshaft (*cam_shaft<1>*). The camshaft was first assembled to the second bushing (*bushing<2>*) and then to the pushrod. *Concentric2* aligns the camshaft and the second bushing. *Coincident4* mates the center plane of the camshaft to that of pushrod, as shown in Figures 12-4f and g. As a result, the camshaft is allowed to rotate about the Z-axis.

(a) Rocker: *Concentric1*   (b) Rocker: *Coincident2*

(c) Pushrod: *Tangent1*   (d) Pushrod: *Distance6*   (e) Pushrod: *Coincident3*

Figure 12-4 Assembly Mates Defined for the Mechanism

(f) Camshaft: *Concentric2*   (g) Camshaft: *Coincident4*   (h) Camshaft: *CamMateTangent1*

(i) Valve: *Parallel1*   (j) Valve: *Distance7*   (k) Valve: *Concentric3*

Figure 12-4 Assembly Mates Defined for the Mechanism (Cont'd)

In addition, the *CamMateTangent1* defines a cam and follower between the cam surface of the camshaft and the cylindrical surface at the bottom of the pushrod, as shown in Figure 12-4h. Note that the assembly should have overall one free degree of freedom at this point. You may either rotate the camshaft, the rocker, or move the pushrod vertically to see the relative motion of the assembly. Use the *Undo* button to restore the assembly to its original configuration. Note that the last mate *Parallel2* listed in the browser was suppressed. This mate was defined to properly orient the camshaft to the *Top* plane of the assembly. This mate is mainly for setting up an initial condition for simulation.

The next and the final part assembled is the valve. The valve was assembled to the first bushing using *Parallel1* and *Distance7* mates, which restrict the valve to move vertically on the *X-Y* plane, as shown in Figures 12-4i and j. The third mate, *Concentric3*, aligns the top cylindrical surface of the valve with the left socket surface of the rocker, as shown in Figure 12-4k. Note that the valve will move vertically along the *Y*-direction, and slightly deviate from the nominal distance of *1.25* in. between its *Right* plane and the *Right* plane of the first bushing. This slight movement is necessary due to the rotation of the rocker. That is why the mate *Parallel1* (between the *Right* plane of the valve and the *Right* plane of the first bushing) is employed instead of a distance mate. You may go over Problem 2 of the exercises at the end of the lesson to learn more about these assembly mates.

Note that this set of assembly mates employed is not the only way to assemble the individual parts for a valid motion model. Is *Concentric3* a desirable mate to connect the valve to the rocker? You may try different ways to mate these parts. As long as the assembly is physically valid, *Motion* is usually able to simulate the motion of a valid assembly that is physically meaningful. An exercise problem (Problem 3) is given at the end of the lesson for you to explore employing a solid body contact, instead of a concentric mate, to connect the valve to the rocker.

Simulation Model

Once we suppress the mate *Parallel2(cam_shaft<1>,Top Plane)*, the mechanism has one free degree of freedom. A rotation on the shaft will be sufficient to uniquely determine the position, velocity, and acceleration of all parts in the mechanism.

The motion model is shown in Figure 12-5, where the *Z*-rotation of the camshaft is driven by a rotary motor with a constant angular velocity of *3,600* degrees/sec, i.e., *600* RPM, about the *Z*-axis of the global coordinate system. In addition, a spring surrounding the valve will be created in order to provide a vertical force to push the rocker up, and therefore, close the valve. The spring has a spring constant of *10* lb$_f$/in and an unstretched length of *2.0* in. The spring is created between the bottom face of the rocker and top face of the valve guide.

Spring

Driver

Figure 12-5 The Simulation Model

12.3 Using *SOLIDWORKS Motion*

Start *SOLIDWORKS* and open assembly file *Lesson12.SLDASM*. Again, always check the units system. Make sure that *IPS* units system is chosen for this example. Suppress mate *Parallel2*.

Click the *Motion Study* tab at the bottom of the graphics area to bring up the *MotionManager* window.

Adding a Rotary Motor

Click the *Motor* button 🔄 from the *Motion* toolbar to bring up the *Motor* dialog box (Figure 12-6). Choose *Rotary Motor* (default). Move the pointer to the graphics area, and pick a circular edge of the camshaft, as shown in Figure 12-7. A circular arrow appears indicating the rotational direction of the rotary motor. A counterclockwise direction (Z-direction) is desired. Choose *Constant speed* and enter *600* RPM (10 revolutions per second) for speed. Click the checkmark ✔ on top of the dialog box to accept the motor definition. You should see a *RotaryMotor1* added to the *MotionManager* tree.

Figure 12-6

Figure 12-7

Defining Spring

Click the *Spring* button 🖥 from the *Motion* toolbar to bring up the *Spring* dialog box. In the *Spring* dialog box (Figure 12-8), the empty field right underneath the *Spring parameters* label is active for you to pick geometric entities to define ends of the spring. Rotate the view and pick the bottom face of the rocker (see Figure 12-9). Rotate the view back, and then pick the top face of the valve guide, as shown in Figure 12-9. Also, a spring should appear in the graphics area, connecting the center points of the two faces.

Figure 12-8 The *Spring* Dialog Box

Figure 12-9 Picking Two Faces to Define the Spring

In case you picked a wrong entity, simply select the entire text in the respective text field in the dialog box, and press the *Delete* key to delete the text. The text field will be cleaned up and ready for you to pick other entities.

In the *Spring* dialog box, enter the following:

Stiffness: 10
Length: 2.0
Coil Diameter: 0.75
Number of coils: 8
Wire Diameter: 0.1

Click *the checkmark* ✔ to accept the definition and close the *Spring* dialog box. A spring node (name: *LinearSpring1*) should appear in the *MotionManager* tree.

Defining and Running Simulation

Choose *Motion Analysis* for the study. Click the *Motion Study Properties* button ⚙ from the *Motion* toolbar. In the *Motion Study Properties* dialog box, enter *500* for *Frames per second*, and click the checkmark on top. Zoom in the timeline area until you can see tenth second marks. Drag the end time key to 0.5 second mark in the timeline area to define the simulation duration.

Click *Calculate* button 🖩 from the *Motion* toolbar to simulate the motion. A 0.5 second simulation will be carried out. After a short moment, you should see the camshaft starts rotating, the pushrod is moving up and down, which drives the rocker and then the valve. The camshaft rotates 5 times in the *0.5-*second simulation duration. You may want to adjust the playback speed to 10% to slow down the animation since the total simulation duration is only 0.5 seconds. The motion looks good.

We will graph the position, velocity, and acceleration of the valve next.

Displaying Simulation Results

Right click the *valve-1* node in the *MotionManager* tree, and choose *Create Motion Plot*. In the *Results* dialog box, choose *Displacement/Velocity/Acceleration*, select *Linear Displacement* and then *Y Component*. Click the checkmark.

A graph like that of Figure 12-10 should appear, where the valve is moving between −2.49 and −2.23 in., traveling about *0.26* in., which is about what was expected, as discussed in Section 12.2.

Figure 12-10 Graph of Valve Position

As shown in Figure 12-10, the flat portion on top indicates that the valve stays completely closed, which spans about *0.066* seconds, approximately *240* degrees of the camshaft rotation in a complete cycle. Therefore, the valve will open for about *0.034* seconds per cycle, roughly *120* degrees.

Graph the *Y*-velocity and *Y*-acceleration of the valve by choosing *Displacement/Velocity/Acceleration*, select *Linear Velocity* and then *Y Component* (and *Linear Acceleration*, then *Y Component*). Click the checkmark. The graphs of the velocity and acceleration are shown in Figures 12-11 and 12-12, respectively. As shown in Figure 12-11, there are two velocity spikes per cycle, representing that the valve is pushed downward (negative velocity) for opening and is being pulled back (positive velocity) for closing, respectively. The valve stays closed with zero velocity.

Figure 12-11 Graph of Valve Velocity

Figure 12-12 Graph of Valve Acceleration

Figure 12-12 reveals high accelerations when the valve is pushed and pulled. Note that such a high acceleration is due to high-speed rotation at the camshaft. This high acceleration could produce large inertial force on the valve, yielding high contact force between the top of the valve and the socket surface in the rocker. We would like to check the reaction force between the top of the valve and the rocker. The graph of the reaction force can be created by expanding the *Mates* node in *MotionManager* tree, right clicking *Concentric3* (between the valve and the rocker) and choosing *Create Motion Plot*. In the *Results* dialog box, choose *Forces*, *Reaction Force*, and then *Y Component*. Click the checkmark. *Motion* shows a warning message, indicating that "this motion study has redundant constraints which can lead to invalid force results. Would you like to replace redundant constraints with bushings..." We will choose *No* for this case since the redundant constraint does not affect this particular result.

The reaction force graph (Figure 12-13) shows that the reaction force between the top of the valve and the socket face of the rocker is about *0.4* lb$_f$, which is insignificant. Note that this small reaction force can be attributed to the small mass of the valve. If you open the valve part and acquire its mass (from pull-down menu, choose *Tools → Evaluate → Mass Properties*), the mass of the valve is *0.02648* lb$_m$. Therefore, the inertia for the valve at the peak accelerations is about *0.02648* lb$_m$×*4,200* in/sec^2 = *111.2* lb$_m$ in/sec^2 = *111.2/386* lb$_f$ = *0.2881* lb$_f$, which is close to the lower peak found in Figure 12-13.

Figure 12-13 Graph of the Reaction Force (*Concentric3*)

If you are not quite sure about why this *386* is factored in for force calculation, please refer to Appendix B for mass and force unit conversions.

Save your model.

Exercises:

1. Redesign the cam by reducing the small arc radius from *0.25* to *0.2* and reducing the center distance
    of the small arc from *0.52* to *0.40*, as shown in Figure E12-1. Repeat the dynamic analysis and check
    reaction force between the valve and the rocker. Does this redesigned cam alter the reaction force?

Figure E12-1 The Cam Profile

2. If we change the *Parallel1* mate between the *Right* plane of the valve and the *Right* plane of the first
    bushing to a distance mate, will the mechanism move? What other changes must be made in order to
    create a valid and movable mechanism similar to what was presented in this lesson?

3. Is the reaction force we obtained from the graph of Figure 12-13 realistic? Will the reaction force
    only involve mass inertia of the valve physically? How about defining a solid body contact, instead
    of a concentric mate, to connect the valve to the rocker? Will the reaction force be more realistic? If
    you do so, what other changes need to be made? (hint: spring, and *Parallel1* mate) Graph the
    reaction force in the *Y*-direction for the valve. Is the reaction force graph different from that of
    Figure 12-13? Explain the differences between these two force results. (hint: show spring force
    graph)

Lesson 13: Simulation of Recreational Waterslide

13.1 Overview of the Lesson

This is the first of the two application lessons offered in this book. The purpose of offering application lessons is to show readers the kind of real-life problems *SOLIDWORKS Motion* is able to support. In *Lesson 13*, we discuss motion study of a riding object on a recreational waterslide. In *Lesson 14*, we present a motion study for a quarter suspension of a Formula SAE style racecar.

Safety and excitement levels of the rider sliding on a waterslide are the two critical criteria for the design of a recreational waterslide. Between the two, safety is far more critical than excitement level. Discovering safety problems after the waterslide is built is usually too late and too costly to correct the problems. Motion study offers a viable alternative in support of the design and analysis of a waterslide in a virtual environment. This alternative would help engineers discover potential safety issues early in the design phase and develop an effective remedy in addressing the design issue.

In this lesson, we start by briefly discussing an analytical calculation method that formulates equations of motion of the riding object and calculates position, velocity, and acceleration of the object on a recreational waterslide[1]. In this method, mathematical representations of a number of commonly employed flume sections are first created in the form of parametric surfaces that are compatible with 3D CAD. These basic building blocks of flume sections have also been implemented in a CAD environment, such as *SOLIDWORKS*. A set of coupled differential equations based on Lagrange's equation that describe the motion of the sliding object are derived, in which friction forces are included. These second order differential equations are then solved using *Mathematica* (www.wolfram.com/mathematica). A real-world waterslide example is presented to illustrate and demonstrate the analytical calculation method. The calculation results, including the trace of the riding object and the overall sliding time, are validated experimentally[2]. In the experimental validation, a scaled waterslide was 3D printed, and the trace of the sliding object was captured by using a high-speed camera. The above discussion should provide readers with a good understanding of the nature of the waterslide simulation. Thereafter, we move on to the discussion of creating a similar motion study using *SOLIDWORKS Motion*. We discuss important modeling considerations and compare the results obtained from *Motion* with those of analytical calculations. We point out a few tasks for future work that may produce simulations capturing physical situations more realistically.

13.2 The Analytical Calculation Method

Mathematic equations of the geometric shape of the three basic flume sections, straight, arc, and curved, as shown in Figure 13-1, are formulated. These flume sections serve as the building blocks for

---

[1] Details can be found in the technical paper: Chang, K.H., Computer-Aided Modeling and Simulation for Recreational Waterslides, Mechanics Based Design of Structures and Machines, An International Journal, 35(3): 229-243, 2007

[2] Details can be found in the technical paper: Chang, K.H. and Chen, C., Design, Prototyping, and Experimental Validation for Recreation Waterslides, Mechanics Based Design of Structures and Machines, 38: 516–529, 2010

composing waterslides. In addition, guard sections (essentially, vertical walls) are added to reinforce a safe design, especially for arc sections. Geometry of all three sections is expressed in parametric surface forms in terms of the parametric coordinates u and w, using CAD geometric dimensions. The general equations of a parametric surface can be expressed as

$$X(u,w) = [X_1(u,w), X_2(u,w), X_3(u,w)]^T \tag{13.1}$$

where $X_i(u,w)$ is the ith coordinate of any given point on the surface of a flume section with prescribed parameters (u,w). Note that these sections will be translated and properly oriented for an overall waterslide configuration by the following translation and rotation operations:

$$X(u,w) = X_o + T(\theta)\,x(u,w) = \begin{bmatrix} X_{0_1} \\ X_{0_2} \\ X_{0_3} \end{bmatrix} + \begin{bmatrix} \cos\theta & 0 & \sin\theta \\ 0 & 1 & 0 \\ -\sin\theta & 0 & \cos\theta \end{bmatrix} \begin{bmatrix} x_1(u,w) \\ x_2(u,w) \\ x_3(u,w) \end{bmatrix} \tag{13.2}$$

(a) The Straight Flume Section

(b) The Arc Flume Section

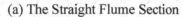

(c) The Curved Flume Section

Figure 13-1 Examples of Standard Flume Sections

where $T(\theta)$ is the rotational matrix that orients the flume section by rotating through angle θ about the X_2 axis, X_{0_i} is the location of the local coordinate system of the section in the waterslide configuration, and $x(u,w)$ is the surface function of the section referring to its local coordinate system x_1-x_2-x_3.

For example, the mathematical representation of a straight section shown in Figure 13-1a can be expressed as,

$$x(u,w) = [-\ell(1-u), h(1-u)+r(1-\sin w\pi), r\cos w\pi]^T, \ u \in [0, 1], \ w \in [0, 1] \tag{13.3}$$

Note that in Eq. 13.3, h and ℓ are the height and length of the straight section, respectively; r is the radius of the cross section (semicircular shape) of the flume section; and w and u are the parametric coordinates along the circumferential and longitudinal directions of the flume section, respectively, as shown in Figure 13-1a. Note that we assume that the cross section of the flume section lies on the x_2-x_3 plane for $u = 1$.

The sliding object is assumed to be a particle with concentrated mass. The well-known Lagrange's equation of motion based on the Hamilton's Principle for particle dynamics[3] was used to derive the equations of motion. The two coupled second order ordinary differential equations that govern the particle motion are obtained as follows:

$$\left.\begin{aligned} k_0\,\ddot{u} &= k_1\,\dot{u}^2 + k_2\,\dot{w}^2 + k_3\,\dot{u}\dot{w} + k_4 \\ k_0\,\ddot{w} &= k_5\,\dot{u}^2 + k_6\,\dot{w}^2 + k_7\,\dot{u}\dot{w} + k_8 \end{aligned}\right\} \tag{13.4}$$

where k_0 to k_8 consist of polynomials of u and w and their products.

Even though the non-friction cases represent the worst-case scenarios in addressing the safety aspect of the waterslide design, adding friction forces into the calculations will provide more realistic simulation results.

For the friction cases, the generalized friction forces $Q = [f_u, f_w]^T$ can be derived as

$$f_u = -\mu\left(gn_2 + a_n \cdot n\right)\left(e_t \cdot X_{,u}\right) \tag{13.5a}$$
$$f_w = -\mu\left(gn_2 + a_n \cdot n\right)\left(e_t \cdot X_{,w}\right) \tag{13.5b}$$

where μ is the friction coefficient; g is the gravitational acceleration; n is the unit normal surface vector; n_2 is the X_2 component of the unit normal vector n; a_n is the normal acceleration of the sliding object; and e_t is the unit vector along the tangential direction of the object's path (see Figure 13-2), which is also the direction of the object velocity X or tangential acceleration a_t. Note that the first term in the first parentheses of Eqs. 13.5a and 13.5b represents the reaction force contributed by the mass of the sliding object. The second term accounts for the reaction force due to acceleration.

The equations of motion for the friction cases are identical to Eq. 13.4, except that k_4 and k_8 are redefined by incorporating the generalized friction forces f_u and f_w.

---

[3] Donald T. Greenwood, *Principles of Dynamics*, Prentice Hall, 1987.

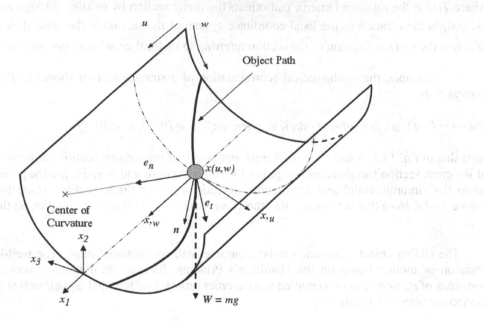

Figure 13-2 Illustration of Unit Vectors for Friction Forces

The initial conditions, including initial position and velocity of the sliding particle, must be provided in order to solve the equations of motion, i.e.,

$$u(0) = u^0, w(0) = w^0, \dot{u}(0) = \dot{u}^0, \text{ and } \dot{w}(0) = \dot{w}^0 \tag{13.6}$$

The system of ordinary differential equations can be solved numerically for positions $u(t)$ and $w(t)$, velocities $\dot{u}(t)$ and $\dot{w}(t)$, and accelerations $\ddot{u}(t)$ and $\ddot{w}(t)$, of the sliding object. These solutions can be further processed for rider positions, velocities, and accelerations in the global Cartesian coordinate system, X_1, X_2, and X_3.

13.3 The Waterslide Example

A waterslide configuration consisting of curved (Section 101), arc with guard (Section 201 and 200/202), and straight sections (Section 301), as shown in Figure 13-3, has been modeled and analyzed. The overall size of the waterslide is about *179* in. × *111* in. × *127* in. Note that the sliding object starts at the center of the cross section ($w = 0.5$) of Section 101, with a small initial velocity ($\dot{u}(0) = 0.01$, or $V = 0.873$ in/sec.). Three cases are assumed: non-friction, small friction of 1% ($\mu = 0.01$) and large friction of 10% ($\mu = 0.1$).

The paths of the sliding object of all three cases can be seen in Figure 13-4a. Note that for the non-friction case, the object runs over to Section 202 briefly, and is very close to the boundary edge of the flume section while entering Section 301 (Figure 13-4b), where $w = 0.919$. Note that when $w \leq 0$ and $w \geq 1$, the object is out of the flume section. Since the non-friction case represents the worst-case scenario, and the object does not run outside of any of the sections, the design is considered safe. Cases 2 and 3 represent friction cases with small and large friction coefficients, respectively. As shown in Figure 13-4,

Case 2 path is very close to that of Case 1 except towards the later part of the slide. This is because a very small friction coefficient is assumed. For Case 3, the path separates more than the first two cases after entering Section 201 due to a larger friction coefficient.

The analytical calculation results are summarized in Table 13-1. The overall sliding times vary from *3.61* to *6.28* seconds for non-friction and large friction cases, respectively. The exit velocity of Cases 2 and 3 also becomes slower due to friction. The same is observed in exit accelerations, varying from *1.45* g to *0.58* g for non-friction and large friction, respectively. Note that the large acceleration of *1.45* g of Case 1 is probably excessive due to the non-friction assumption.

Figure 13-3 A Waterslide Consisting of Three Flume Sections

(a)   (b)

Figure 13-4 Object Paths on the Waterslide

Table 13.1 Results Obtained from Analytical Calculations

Case	Description	Total Sliding Time (Second)	Exit Velocity (Inch/Sec, MPH)	Exit Acceleration (Inch/Sec^2, g)
1	Non Friction	*3.608*	*501, 28.5*	*561, 1.45*
2	Small Friction ($\mu = 0.01$)	*4.557*	*260, 14.8*	*255, 0.66*
3	Large Friction ($\mu = 0.1$)	*6.275*	*162, 9.2*	*225, 0.58*

For designing waterslides, the non-friction cases set boundary for safety considerations. However, their accelerations are usually excessive, and therefore, not suitable for the designs that involve excitement levels, which are largely determined by accelerations.

13.4 Experimental Validation

In this section, we briefly discuss an experimental method employed for validating the design of recreational waterslides obtained from analytical calculations (or simulations). A scaled physical prototype of the waterslide designed in CAD is first fabricated using a 3D printer, as shown in Figure 13-5. The surface of the printed waterslide is polished and treated to reduce friction, which plays a key role in this study. The object sliding path and total sliding time are then captured using a high-speed digital camera (see Figure 13-6).

A semi-spherical steel ball of radius *0.078* in. is employed as the sliding object for the experiments. The semi-spherical ball is chosen mainly for the purpose of preventing rolling. The object mimics closely the situation of sliding with point contact, which is close to the assumption made in the analytical calculations.

(a) CAD Model   (b) STL Model   (c) Physical Model

Figure 13-5 Waterslide Solid Model for 3D Printing

(a) Physical Setup   (b) Schematic Drawing

Figure 13-6 Experimental Setup

Table 13-2 Comparison of Sliding Times (Seconds) between Analytical Calculations (Sim.) and Experiments (Exp.)

	Sim. μ =0.15	(Exp.-Sim.)/Sim.	Sim. μ =0.155	(Exp.-Sim.)/Sim.	Sim. μ =0.16	(Exp.-Sim.)/Sim.
w=0.5	1.836	+1.9%	1.910	-2.1%	1.996	-6.3%
w=0.25	1.796	+1.2%	1.867	-2.7%	1.945	-6.6%
w=0.75	1.768	+3.9%	1.834	+1.6%	1.913	-3.9%

(a) First scenario $w = 0.5$

(b) Second scenario $w = 0.25$

(c) Third scenario $w = 0.75$

Figure 13-7 Object Path Comparison from Top View

The friction coefficient between the sliding object and the waterslide surface of the physical model (3D printed) was measured experimentally to be between $\mu = 0.15$ and 0.16. Three test scenarios were carried out for respective starting positions, i.e., $w = 0.5, 0.25$, and 0.75. Overall sliding times and object paths were compared.

The comparison of overall sliding times between the experiments and analytical calculations is summarized in Table 13-2. Note that analytical calculations were conducted for friction coefficients 0.15, 0.155, and 0.16. From Table 13-2, the comparison shows the overall sliding time obtained from experiments agree with those of analytical calculations extremely well. For $\mu = 0.155$, the maximum difference is 2.7%. The largest difference of all cases is 6.6% for $w = 0.5$ and $\mu = 0.16$, indicating that the overall sliding time agrees well between experiments and analytical calculations.

Similar to the overall sliding time, sliding paths obtained from the experiments agree well with those of analytical calculations, as shown in Figure 13-7. Analytical calculation results of both $\mu = 0.15$ and 0.16 are compared with those of experiments. Figure 13-7 shows the sliding path from top view for all three test scenarios. The sliding paths show, in general, excellent agreement between experiments and analytical calculations. There are a few minor differences in path as circled in Figure 13-7. For example, the object is slightly further away from the center line of Section 201 in analytical calculations compared with that of the experiment. This deviation appears more in the scenario of $w = 0.25$, as shown in Figure 13-7b.

Overall, the analytical calculation (or simulation) results agree with those of experiments. The overall sliding times recorded in experiments fall between +3.9% and −6.6% of those analytical calculations where μ is between 0.15 and 0.16. The sliding paths in Section 101 and 201 agree well between experiments and analytical calculations. Some path deviation in Section 301 is observed. The deviation accumulates and causes more separation when the object slides closer to the exit of the final flume surface.

13.5 The Motion Study Model

The same waterslide example shown in Figure 13-3 has been created in *SOLIDWORKS*. The assembly model consists of a sliding object of semi-spherical shape and a waterslide identical to that of Figure 13-3. Motion study was carried out for three cases, no friction, small friction ($\mu = 0.01$), and large friction ($\mu = 0.1$). Results obtained from *Motion* are compared with those obtained from analytical calculations. The sliding object is a semi-sphere of 0.025in. in diameter. The semi-spherical ball is chosen mainly for the purpose of preventing rolling. A small diameter is assumed to comply with the assumption of concentrative mass.

SOLIDWORKS Part and Assembly Files

There are three model files created for this lesson: *R50_Slide.SLDPRT*, *Rider-Semi-Sphere-r025.SLDPRT*, and *Waterslide-Semi-Sphere r025.SLDASM*. The part file *R50_Slide.SLDPRT* consists of three sweep solid features for the respective flume sections; Sections 101, 200/201/202, and 301, as shown in Figure 13-8. In addition, position of the riding object obtained from analytical calculations; that is, those shown in Figure 13-4 for the respective three cases (no friction, small friction: mu001, and large friction: mu01), are imported into the *SOLIDWORKS* part model, *R50_Slide.SLDPRT*, as curves (by choosing from the pull-down menu *Insert → Curve → Curve Through XYZ Points*). These curves are shown in Figure 13-9a, b, c, for Cases 1, 2, and 3, respectively.

Figure 13-8 The Sweep Solid Features of the Respective Flume Sections

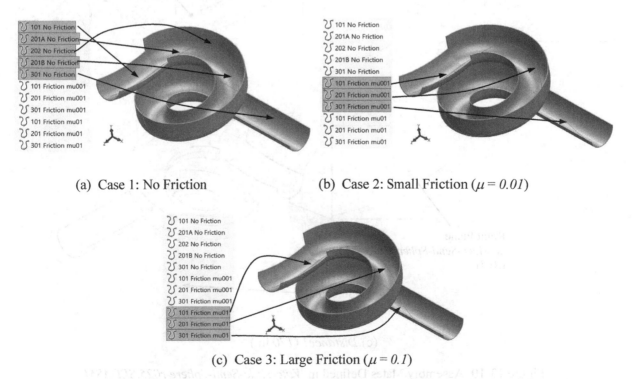

(a) Case 1: No Friction

(b) Case 2: Small Friction ($\mu = 0.01$)

(c) Case 3: Large Friction ($\mu = 0.1$)

Figure 13-9 Paths of the Sliding Object Obtained from Analytical Calculations

The sliding object or rider, *Rider-Semi-Sphere-r025.SLDPRT*, consists of a revolve solid feature of semi-sphere with diameter *0.025*in. The rider is assembled to the waterslide by using three assembly mates: *Tangent1*, *Coincident1*, and *Distance1*, as shown in Figure 13-10a, b, and c, respectively. Note that in *Motion* analysis, all three mates *Tangent1*, *Coincident1* and *Distance1* are suppressed.

Motion Simulation Model

Three motion studies, *No Friction*, *mu001*, and *mu01*, have been created, corresponding to the respective three cases of the analytical calculations.

A gravitational acceleration *–386* in/sec^2 is defined in the *Y*-direction of the global coordinate system.

A solid body contact is added to characterize the characteristics of contact between the rider and the surface of the waterslide. A static friction coefficient μ_k is defined to simulate the three cases, i.e., $\mu_k = 0$, $\mu_k = 0.01$, and $\mu_k = 0.1$, respectively. The friction coefficient is entered through the *Contact* dialog box, as circled in Figure 13-11.

(a) *Tangent1*   (b) *Coincident1*

(c) *Distance1* (*170* in.)

Figure 13-10 Assembly Mates Defined in *Waterslide-Semi-Sphere r025.SLDASM*

It is important to note that the characteristics of the contact condition are also determined by the elastic properties circled in the *Contact* dialog box (see Figure 13-11). In the example, we selected *Impact* and stayed with the default impact properties, including:

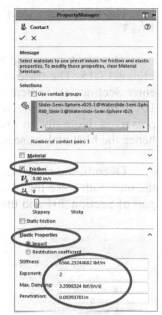

Stiffness *k*: Sets the stiffness to approximate the material stiffness at the boundary of interaction between the two parts in collision.

Exponent *e*: Sets the exponent in the assumed exponential force versus displacement model. In general, for a stiffening spring, *e > 1.0*. For a softening spring, *0 < e < 1.0*.

Max. damping c_{max}: Sets the maximum damping coefficient of the boundary interaction.

Penetration d_{max}: Sets the positive boundary penetration value at which the *SOLIDWORKS Motion* solver applies c_{max}, the Max. damping value.

Figure 13-11

The *SOLIDWORKS Motion* solver uses a damping value that is less than Max. damping c_{max} when calculating penetrations less than Penetration d_{max}.

The impact force F_n is a function of displacement and velocity, calculated by an internal function of *SOLIDWORKS* as:

$$F_n = k\,(h^e) + \text{STEP}\,(h, 0, 0, d_{max}, c_{max})\,dh/dt \tag{13.7}$$

where *h* is the penetration of one geometry into another, and *dh/dt* is the penetration velocity at the contact point, and *STEP* is a step function that returns a smooth step function of the damping coefficients depending on the severity of the penetration. Note that if the penetration *h* is zero, i.e., the objects are not in contact, the impact force is set to be zero.

13.6 Using *SOLIDWORKS Motion*

Start *SOLIDWORKS* and open *Waterslide-Semi-Sphere r025.SLDASM*. You should see *Waterslide-Semi-Sphere r025.SLDASM* appears in the graphics area like that of Figure 13-8. The rider solid part, *Semi-Sphere r025.SLDPRT*, is too small to see. You may zoom into the left end of Section 101 to see it. You may also turn on the display for *Origin* to locate the rider by choosing from the pull-down menu: *View → Hide/Show → Origin*.

Motion studies for all three cases have been completed for you. You may click any motion study tab at the bottom of the graphics window to review the motion study results. For example, we click the *No Friction* tab. We display the trace of the rider obtained from *Motion* by right clicking from the *MotionManager* tree *Rider Trace* under *Results*, and select *Show Plot* (see Figure 13-12). You may click the *Play from Start* button ▶ (circled in Figure 13-12) to play the rider motion of the 4.5-second simulation. The trace of the rider obtained from *Motion* study should appear (in black color) like that of Figure 13-13.

Next, we select from the *FeatureManager* tree, *101 No Friction*, *201A No Friction*, *202 No Friction*, *201B No Friction*, and *301 No Friction*, and right click to choose *Show* button ◉ (circled in Figure 13-

14). The path of the rider obtained from analytical calculations appears in blue color like that of Figure 13-13.

The black trace and blue path agree well, except for the last part of Section 201, as pointed out in Figure 13-15. The traces become more separated in Section 301. A closer look on the traces as the rider enter Section 201 indicates that the rider flies off the flume surface (black trace) as it exits from Section 101. This is because the junction across Sections 101 and 201 is not smooth, in which C^1-continuity is not preserved. Note that in the analytical calculation, the rider is assumed to stay right on the flume surface, hence the blue path as seen in, for example, Figure 13-15.

Overall, the rider trace of motion study (black trace) indicates that the rider is safer compared with the blue path since the black trace does not reach the outer edge of the flume section, for example the later part of Section 201 and the beginning part of Section 301 like that of the blue path.

Figure 13-12 Selection to Show Rider Trace

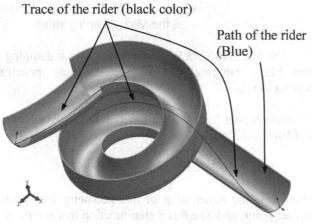

Figure 13-13 Trace of the Rider, Obtained from *Motion* in Black, Path Obtained from Analytical Calculations in Blue

Figure 13-14

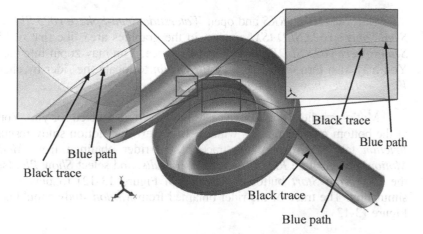

Figure 13-15 A Closer-Look on the Black Trace and Blue Path

The overall sliding time is *4.462* seconds. At the exit, the linear velocity of the rider is *272.57* in/sec in magnitude. The sliding time can be obtained by showing the plot *X-Position*, and exporting the position data to a spreadsheet. The *X*-position of the rider at the beginning is *–170* in., hence the *X*-position of the rider exiting the slide is *0*. The spreadsheet data indicates (see Figure 13-16) when the rider reaches *0* (in fact, *–0.02706*, closest to *0*), the time is *4.462* seconds. The velocity of the rider at the exit of the slide is obtained by showing *Rider Velocity* plot, exporting the velocity data into a spreadsheet, and checking the velocity at the time *4.462* seconds.

Next, we select the *mu001* tab, and repeat the steps discussed above.

We show the traces of the rider obtained from motion study (black curves) and analytical calculation (blue curves) like that of Figure 13-17a. The black and blue traces agree well for the entire simulation. The overall sliding time is *4.599* seconds; at the time, the linear velocity in magnitude is *259.31* in/sec.

We do the same for the large friction case *mu01*.

We show the traces of the rider obtained from motion study (black curves) and analytical calculation (blue curves) like that of Figure 13-17b. The black and blue traces agree well for the most part, except for a slight difference observed in Section 301. The overall sliding time is *6.422* seconds, at the time, the linear velocity in magnitude is *132.56* in/sec.

	A	B
991	4.391365	-19.2444
992	4.395971	-18.0021
993	4.4	-16.9151
994	4.41	-14.215
995	4.42	-11.5128
996	4.421159	-11.1995
997	4.43	-8.80879
998	4.437192	-6.86298
999	4.44	-6.10316
1000	4.45	-3.39585
1001	4.454442	-2.19278
1002	4.46	-0.68672
1003	4.462434	-0.02706
1004	4.465123	0.701929
1005	4.47	2.023736
1006	4.48	4.734247

Figure 13-16

(a) Small Friction (mu001)   (b) Large Friction (mu01)

Figure 13-17 Trace of the Rider, Obtained from *Motion* in Black, Obtained from Analytical Calculation in Blue

Table 13.3 Comparison Between Results Obtained from Motion Study and Analytical Calculations

Case	Description	Total Sliding Time (Second)		Exit Velocity (in/sec)	
		Analytical Calculations	Motion Study	Analytical Calculations	Motion Study
1	Non Friction	3.608	4.462	501	272.57
2	Small Friction ($\mu = 0.01$)	4.557	4.599	260	259.31
3	Large Friction ($\mu = 0.1$)	6.275	6.422	162	132.56

We compare the results obtained from *Motion* and analytical calculations. As shown in Table 13-3, the overall sliding time and exit velocity of Case 1: non friction, are not consistent. Motion study results show a longer sliding time and significantly smaller exit velocity. This could be because of the elastic properties existing in defining the contact condition. A completely friction free simulation may not have been created in using *Motion*. Results of Case 2: small friction agree very well. For Case 3: large friction, sliding times are close, exit velocities are not identical but close.

Which results are more accurate, motion study or analytical calculations? Both are accurate for a small sliding object, except for the non-friction case explained before. However, in real-life scenarios, riders are human that cannot be assumed as a concentrative mass. The simulation results, either obtained from *Motion* or analytical calculations, only offer an estimate for the real-life scenarios when a human is riding the waterslide. More research and development must be carried out to bringing a realistic human model into the computer so that a more realistic simulation can be carried out.

Lesson 14: Kinematic Analysis of a Racecar Suspension

14.1 Overview of the Lesson

In this lesson, we take a quarter of a racecar suspension and create a motion model for kinematic analysis. The racecar model employed, as shown in Figure 14-1, is a Formula SAE (Society of Automotive Engineers) style racecar designed and built by engineering students at the University of Oklahoma (OU) during 2005-2006. Each year engineering students throughout the world design and build formula-style racecars and participate in the annual Formula SAE competitions (www.sae.org/students). The result is a great experience for young engineers working on a meaningful engineering project as well as an opportunity to work in a dedicated team environment.

The suspension of the entire racecar was modeled for both kinematic and dynamic analyses. These analysis results were validated using experimental data. The experimental data were acquired by mounting a data acquisition system on the racecar and driving the racecar on a test track following specific driving scenarios that are consistent with those of the simulations. These results were used to aid the suspension design for handling and cornering. Assembling an entire vehicle suspension for motion analysis is non-trivial and is beyond the scope of this book. Therefore, only the right front quarter of the racecar suspension, as shown in Figure 14-2, will be employed in this lesson. The purpose of this lesson is mainly to show you that *Motion* is capable of supporting the design of kinematic characteristics of the vehicle suspension system. Instead of repeating the detailed steps of constructing the motion model in *SOLIDWORKS*, in this lesson, we will start with an assembled motion model. The only component we will add to the motion model is the road profile. The road profile is characterized by the geometric shape of a profile cam, which will be assembled to the tire using a cam-follower assembly mate in *SOLIDWORKS*.

(a) Manufactured Racecar on Display

(b) Racecar Designed in *Creo*

Figure 14-1 Formula SAE Racecar Designed and Built by OU Engineering Students

14.2 The Quarter Suspension

Physical Model

The quarter suspension consists of major components that essentially define the kinematic and dynamic characteristics of the racecar. These components include upper and lower control arms, upright, rocker, shock, push rod, tie rod, and wheel and tire, as shown in Figure 14-3. The dangling end of the shock, both control arms, rocker, and tie rod are connected to the chassis frame using numerous joints. The chassis frame is assumed fixed and the tire is pushed and pulled by the profile cam (not shown) mimicking the road profile. Two saved views, *View A* and *View B,* shown in Figure 14-3, are created in the assembled model and will be used for illustrations throughout this lesson.

Figure 14-2 The Right Front Quarter of the Racecar Suspension (View A)

(a) Saved View: View B   (b) Saved View: View A

Figure 14-3 Major Components of the Quarter Suspension

The tire of the quarter suspension will be in contact with the profile cam representing the road profile. As shown in Figure 14-4, the geometry of the cam consists of two circular arcs of radius 6.65 in. (AB and FG), which are concentric with the cam center. Therefore, when the cam rotates, these two circular arcs do not push or pull the tire, resulting in two flat segments of the road profile, as shown in Figure 14-5. In addition, the circular arc CDE is centered 4 in. above the cam center with a radius of 4 in. Therefore, when the cam rotates, arc CDE pushes the tire up, mimicking a hump of 1.35 in. (that is 1.35 = 8–6.65, peak at point D). A ditch is characterized by an 8 in. arc (HIJ) centered at 3 in. above the cam center. As the cam rotates, arc HIJ creates a ditch of 1.65 in. deep—that is 1.65 = 6.65–(8–3). The remaining straight lines and arcs provide smooth transitions among flats, humps, and ditches in the road profile.

Figure 14-4 Geometry of the Profile Cam

Figure 14-5 Road Profile Generated by the Profile Cam

Based on the geometry of the profile cam, this quarter suspension will go over a 1.35 in. hump and a 1.65 in. ditch in one complete rotation of the profile cam. Note that since the radius of arc AB is 6.65 in., the cam will cause the quarter suspension to travel roughly 41.8 in. (3.48 ft.) in one complete rotation. Since the profile cam will rotate a complete cycle in one second, the suspension travels about 3.48 ft/sec, i.e., 2.37 MPH, a very slow motion.

Note that the in-lb$_f$-sec unit system has been employed for all parts and assemblies. Plain carbon steel is assumed for all parts.

SOLIDWORKS Part and Assembly Files

All *SOLIDWORKS* parts and assemblies are provided for this lesson. In addition, all parts and subassemblies are assembled, except for the profile cam (part name: *profile.sldprt*). The profile cam will be assembled to the tire using a cam-follower connection we learned in Lesson 12. A rotary motor will be added to rotate the cam at a constant angular velocity of 360 degrees/sec, therefore pushing and pulling the tire along the vertical direction, mimicking the situation where the racecar goes over humps and ditches.

Note that the example files you downloaded from the publisher's web site should consist of 23 files: 17 parts, and 6 assemblies, as listed in Table 14-1. The quarter suspension *quarter_suspension.sldasm* is completely assembled except for the road profile (*profile.sldprt*). We will start with this assembly and bring in the profile cam, which will be assembled to the tire using a cam-follower connection. In addition, a completely assembled motion model, *quarter_suspensionWithResults.sldasm*, is included for your reference. You may want to open this motion model and bring up this result file to preview the motion animation of the quarter suspension system.

Table 14-1 List of Files in Lesson 14 Folder

Assembly	Part/Subassemblies			Remarks
quarter_suspension.sldasm				Assembly to start the lesson
	hardpoints.sldprt			Datum features
	fr_rocker_r1.sldasm			Rocker
		Rocker_Redesign.sldprt		
		Bearing.sldprt (2)		
	shock_upper.sldprt			Shocks
	shock_lower.sldprt			
	lca.sldprt			Lower Control Arm
	prod.sldasm			Push Rod
		prod_Redesign_L.sldprt		
		prod_Redesign_M.sldprt		
		prod_Redesign_S.sldprt		
	wheel_r1.sldasm			Wheel
		wheel.sldprt		
		fr_upright_r1.asm		
			RH_upright	
			RH_upright_slim	
			RH_steer_arm	
	uca.sldprt			Upper Control Arm
	trod.sldprt			Tie Rod
	wheel_ref.sldprt			Reference part
	profile.sldprt			To be assembled
quarter_suspensionWithResults.sldasm				Complete Motion Model

Figure 14-6 Assembly Mates Listed in the Browser

Figure 14-7 The Reference Geometric Entities (or Datum Features) Included in *HARDPOINTS* Employed for Defining Assembly Mates (View B)

There are nineteen assembly mates, including sixteen coincident, one concentric, one angle, and one parallel, as listed in the browser (see Figure 14-6). You may want to expand the *Mates* node in *SOLIDWORKS* browser to review the assembly mates. Click any of the mates; you should see the entities selected for the assembly mate highlighted in the graphics area.

The first component, *HARDPOINTS*, shown in the browser consists of nothing but datum features (or reference geometric entities) and is anchored to the assembly. As a result, *HARDPOINTS* is completely fixed to the assembly. As shown in Figure 14-7, there are three planes (*GROUND*, *CENTERLINE*, and *WHEEL_AXIS*), nine 3D axes (*UCA_F_*, *UCA_R_*, *LCA_F_*, *LCA_R_*, *ROCKET_PIVOT_*, *ROCKER_AXIS_*, *SHOCKER_FRAME_TROD_INNER_*, *ROCKER_PROD_REF_*, and *A_1_*), nine points (*UCA_F*, *UCA_R*, *LCA_F*, *LCA_R*, *ROCKET_PIVOT*, *ROCKER_AXIS*, *SHOCKER_FRAME*, *TROD_INNER*, and *ROCKER_PROD_REF*), and four axes (*LCA_AXIS*, *UCA_AXIS*, *A_1*, and *ROCKER_PIVOT_AXIS*) defined in *HARDPOINTS* for support of defining assembly mates.

(a) *fr_rocker_r1: Coincident2 (Axes)*

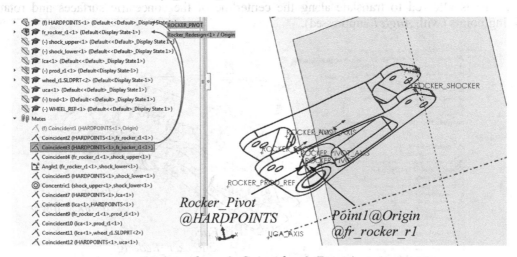

(b) *fr_rocker_r1: Coincident3 (Points)*

Figure 14-8 *fr_rocker_r1* assembled to *HARDPOINTS* (View B)

The second component, *fr_rocker_r1*, is assembled to *HARDPOINTS* by coinciding two axes defined in *Coincident2* (*Rocker_Pivot_axis@fr_rocker_r1* and *Rocker_Pivot_axis@HARDPOINTS*) and coinciding two points defined in *Coincident3* (*Point1@Origin@fr_rocker_r1* and *Rocker_Pivot@HARDPOINTS*), as shown in Figure 14-8. As a result, *fr_rocker_r1* is allowed to rotate along the common axis, which is part of the *HARDPOINTS* and is stationary.

The next component, *shock_upper*, is assembled to *fr_rocker_r1* by coinciding two points, defined in *Coincident4* (*Shocker_rocker@shock_upper* and *Rocker_Shocker@fr_rocker_r1*), as shown in Figure 14-9. As a result, *shock_upper* is allowed to rotate in all three directions at the coincident point.

Figure 14-9 *shock_upper* assembled to *fr_rocker_r1* (View B)

Next, *shock_lower* is assembled to the rocker by specifying a 65 degree angle between the two flat faces on the *rocker* and *shock_lower*, respectively, as shown in Figure 14-10a. Note that this angle mate defines the initial configuration of the suspension. Before running simulation, this mate will have to be suppressed. In addition, *shock_lower* is assembled to *HARDPOINTS* by coinciding two points defined in *Coincident5* (*Shocker_frame@shock_lower* and *Shocker_frame@HARDPOINTS*), and assembled to *shock_upper* by concentrating two cylindrical surfaces defined in *Concentric1* (*Face<1>@shock_upper* and *Face<2>@shock_lower*), as shown in Figure 14-10b and 14-10c, respectively. As a result, *shock_lower* is allowed to translate along the centerline of the concentric surfaces and rotate at the coinciding points (with *Angle1* suppressed).

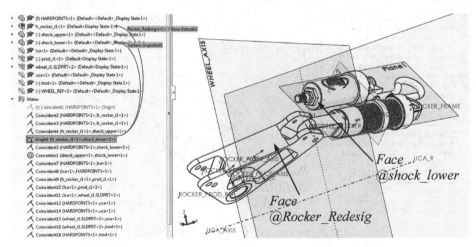

(a) *shock_lower* assembled to *Rocker*: *Angle1 (Faces)*

Figure 14-10 *shock_lower* assembled

(b) *shock_lower* assembled to *HARDPOINTS*: *Coincident5 (Points)*

(c) *shock_lower* assembled to *shock_upper*: *Concentric1 (Cylindrical Surfaces)*

Figure 14-10 *shock_lower* assembled (View B) (cont'd)

Next, *lca* (lower control arm) is assembled to *HARDPOINTS* by coinciding two axes defined in *Coincident7* (*Lca_axis@lca* and *Lca_axis@HARDPOINTS*), and coinciding two points defined in *Coincident8* (*Lca_fr@lca* and *Lca_f@HARDPOINTS*), as shown in Figure 14-11a and 14-11b, respectively. As a result, *lca* is allowed to rotate along the common axes.

prod (push rod) is assembled to *fr_rocker_r1* by coinciding two points defined in *Coincident9* (*Prod_inner@prod* and *Rocker_prod@fr_rocker_r1*), and assembled to *lca* by coinciding two points defined in *Coincident10* (*Prod_outer@prod* and *Lca_prod@lca*), as shown in Figure 14-12a and 14-12b, respectively. As a result, *prod* is allowed to rotate between these two sets of coincident points.

wheel_r1 is temporarily assembled to *lca* by coinciding two points defined in *Coincident11* (*Upright_lca@wheel_r1* and *Point1@Origin@lca*), as shown in Figure 14-13. As a result, *wheel_r1* is allowed to rotate in all three directions at the coincident points. More mates will be added to assemble the wheel later.

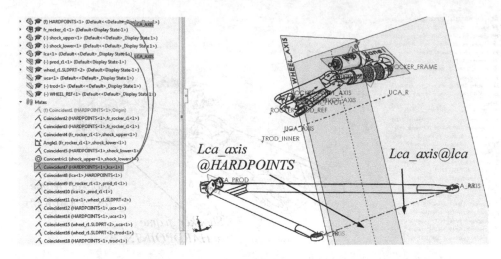

(a) *lca* assembled to *HARDPOINTS*: *Coincident7 (Axes)*

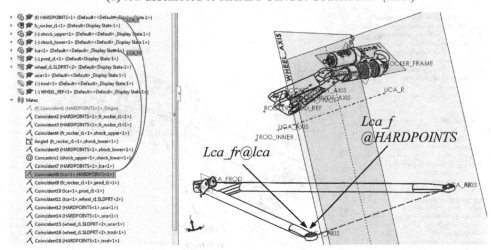

(b) *lca* assembled to *HARDPOINTS*: *Coincident8 (Points)*

Figure 14-11 *lca* assembled (View B)

uca (upper control arm) is assembled to *HARDPOINTS* by coinciding two axes defined in *Coincident12* (*Ucr_axis@uca* and *Uca_axis@HARDPOINTS*), two points defined in *Coincident14* (*Ucr_fr@uca* and *Uca_f@HARDPOINTS*), and as shown in Figure 14-14a and 14-14b, respectively. As a result, *uca* is allowed to rotate along the common axes. In addition, *wheel_r1* is assembled to *uca* by coinciding two points defined in *Coincident15* (*Upright_uca@wheel_r1* and *Point1@Origin@uca*), as shown in Figure 14-14c.

Next is the tie rod. *trod* is assembled to *wheel_r1* by coinciding two points defined in *Coincident16* (*Trod_upright@wheel_r1* and *Trod_outer@trod*), as shown in Figure 14-15a. Moreover, the tie rod is assembled to the *HARDPOINTS* by coinciding two points defined in *Coincident16* (*Trod_inner@trod* and *Trod_inner@HARDPOINTS*), as shown in Figure 14-15b. As a result, the *trod* is allowed to rotate in all three directions at the coincident points.

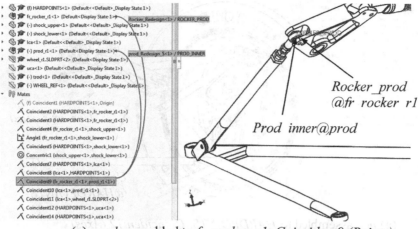

(a) *prod* assembled to *fr_rocker_r1*: *Coincident9 (Points)*

(b) *prod* assembled to *lca*: *Coincident10 (Points)*

Figure 14-12 *prod* assembled (View B)

Figure 14-13 *wheel_r1* temporarily assembled to *lca* (View B)

(a) *uca* assembled to *HARDPOINTS: Coincident12 (Axes)*

(b) *uca* assembled to *HARDPOINTS: Coincident14 (Points)*

(c) *wheel_r1* assembled to *uca: Coincident15*

Figure 14-14 *uca* assembled (View B)

(a) *trod* assembled to *wheel_r1: Coincident16 (Points)*

(b) *trod* assembled to *HARDPOINTS: Coincident18 (Points)*

Figure 14-15 *trod* assembled (View B)

Now, we discuss two more assembly mates for the *wheel_ref*. Note that *wheel_ref* is a part that is created as a patch to aid in defining a cam and follower mate between the wheel and the road profile to be discussed shortly. The *wheel_ref* is assembled to *wheel_r1* by coinciding two points defined in *Coincident21* (*wheel_center@wheel_ref* and *Point1@Origin@wheel_r1*), and assembled to the *Top Plane* by keeping a face and the *Top Plane* in parallel defined in *Parallel2* (*Face<1>@wheel_ref* and *Top Plane*), and as shown in Figure 14-16a and 14-16b, respectively.

After all the nineteen assembly mates are defined, the quarter suspension assembly is completely assembled as shown in Figure 14-17 with the standard three views and a saved view (View B). Note that the assembly is fully constrained. You will have to suppress *Angle1* to allow for motion between components in the suspension assembly. If you suppress *Angle1*, click the wheel and drag it up and down; you will see the movements of individual parts. You may choose from the pull-down menu *Edit > Undo Move Component* to bring the assembly to its original configuration.

(a) *wheel_ref* assembled to *wheel_r1: Coincident21 (Points)*

(b) *wheel_ref* assembled to *Top Plane: Parallel2 (Planes)*

Figure 14-16 *wheel_ref* assembled (View B)

Simulation Model

From motion simulation perspective, there are nine bodies defined in this motion model, including the ground body. All the key datum features, including datum coordinate systems, datum axes, and datum points, required for assembly are collected in *HARDPOINTS*, as shown in Figure 14-7. The *HARDPOINTS* was assembled to the assembly *quarter_suspension*. Therefore, *HARDPOINTS* belongs to the ground body. Note that the *X*-axis of the reference triad points to the backward direction of the suspension, as shown in Figure 14-7, and the wheel is about 2.4 in. above the ground in the *Z*-direction (that is the distance between the lowest point in the tire and the reference plane *GROUND*, as shown in Figure 14-18). Note that the distance is measured by choosing *Tools > Evaluate > Measure* from the pull-down menu.

Figure 14-17 The Quarter Suspension Assembly in Standard Three Views and a Prescribed View B
(upper right)

Edge<1>@wheel_r1

GROUND

Figure 14-18 A 2.4 in. Distance Between the Lowest Point of Wheel and the *GROUND (*Origin of the
quarter_suspension Assembly*)*

The assembly mates are converted internally to fourteen kinematic joints, as shown in Figure 14-19. These joints are listed in Table 14-2 together with the nine bodies and their corresponding assembly mates.

Figure 14-19 Kinematic Joints Converted from the Assembly Mates (View B)

Table 14-2 Kinematic Joints and Assembly Mates

Body	Part or Assembly	Joints	Assembled to	Assembly Mates
Ground Body	*hardpoints.sldprt*	Rigid	*quarter_suspension*	
Body1	*ft_rocker_r1.sldasm*	Pin (1)	*hardpoints*	*Coincident2*, axes: *Rocker_Pivot_axis@fr_rocker_r1* and *Rocker_Pivot_axis@HARDPOINTS* *Coincident3*, points: *Point1@Origin@fr_rocker_r1* and *Rocker_Pivot@HARDPOINTS*
Body2	*shock_upper.sldprt*	Ball (1)	*ft_rocker_r1*	*Coincident4*, points: *Shocker_rocker@shock_upper* and *Rocker_Shocker@fr_rocker_r1*
Body3	*shock_lower.sldprt*	Ball (2)	*hardpoints*	*Coincident5*, points: *Shocker_frame@shock_lower* and *Shocker_frame@HARDPOINTS*
		Cylinder (1)	*shock_upper*	*Concentric1*, cylindrical surfaces: *Face<1>@shock_upper* and *Face<2>@shock_lower*
Body4	*lca.sldasm*	Pin (2)	*hardpoints*	*Coincident7*, axes: *Lca_axis@lca* and *Lca_axis@HARDPOINTS* *Coincident8*, points: *Lca_fr@lca* and *Lca_f@HARDPOINTS*
Body5	*prod.sldasm*	Ball (3)	*lca*	*Coincident10*, points: *Prod_outer@prod* and *Lca_prod@lca*
		Ball (4)	*fr_rocker_r1*	*Coincident9*, points: *Prod_inner@prod* and *Rocker_prod@fr_rocker_r1*

Table 14-2 Kinematic Joints and Assembly Mates (cont'd)

Body	Part or Assembly	Joints	Assembled to	Assembly Mates
Body6	*wheel_r1.sldasm*	Ball (5)	*lca*	*Coincident11*, points: *Upright_lca@wheel_r1* and *Point1@Origin@lca*
		Ball (6)	*uca*	*Coincident15*, points: *Upright_uca@wheel_r1* and *Point1@Origin@uca*
		Ball (7)	*trod*	*Coincident16*, points: *Trod_upright@wheel_r1* and *Trod_outer@trod*
Body7	*uca.sldasm*	Pin (3)	*hardpoints*	*Coincident12*, axes: *Ucr_axis@uca* and *Uca_axis@HARDPOINTS* *Coincident14*, points: *Ucr_fr@uca* and *Uca_f@HARDPOINTS*
Body8	*trod.sldasm*	Ball (8)	*hardpoints*	*Coincident18*, points: *Trod_inner@trod* and *Trod_inner@HARDPOINTS*

14.3 Using *SOLIDWORKS Motion*

Assembling the Profile Cam

Start *SOLIDWORKS* and open *quarter_suspension.sldasm*. You should see *quarter_suspension.sldasm* appear in the graphics area with the default view *View B* (see Figure 14-3a). Note that the *quarter_suspension.sldasm* consists of nine components, listed in the browser, i.e., *HARDPOINTS*, *fr_rocker_r1*, *shock_upper*, *shock_lower*, *lca*, *prod*, *wheel_r1.sldasm*, *uca*, and *trod.sldasm*. You may choose the *Configuration* tab 🔲 above the browser, right click *ExpView1* and choose *Explode* to see the explode view of the assembly, as shown in Figure 14-20. Right click *ExpView1* and choose *Collapse* to restore the unexplode view of the assembly. Check the units system and make sure that *IPS* units system is chosen for this example.

Figure 14-20 Displaying the Explode View Under the *Configuration* Tab

To add the cam to the assembly, click *Insert > Component > Existing Part/Assembly* from the pull-down menu and click *Browse* in *Insert Component* dialog box to insert the part *profile*. Place the profile cam at an arbitrary location in the graphics area. Click *Insert > Mate* from the pull-down menu to bring up the *Mate* dialog box (Figure 14-21). Pick reference axis *A_1* (*profile*) and reference axis

A_1(HARDPOINTS) from the feature tree (Figure 14-22). Make sure the mate type *Coincide* ⚔ is selected, and then click ✔ on top of the dialog box to accept the mate.

Pick the *Front Plane* in *profile* and the *CENTERLINE* plane in *HARDPOINTS*, then click the *Distance* ⊢⊣ button in the *Mate* dialog box. Enter *24.5* and press the *Enter* key (See Figure 14-23). You may need to check or cancel the *Flip dimension* option to obtain the correct mate as shown in Figure 14-23. Click the checkmark ✔ to accept the definition.

Figure 14-21 The *Mate* Dialog Box

Figure 14-22 Picking the Axes for Assembly

Figure 14-23 Picking Planes for Defining a Distance Mate

In order to fix the rotation degree of freedom of the cam, first pick the *Top Plane* of *profile* and the *GROUND* plane in *HARDPOINTS*, as shown in Figure 14-24. The *Parallel* mate is automatically selected, and the orientation of the cam should be similar to that shown in Figure 14-24, which means we want the hump of the profile cam to point downward at the beginning of the simulation. Click the checkmark ✔ twice to close the *Mate* dialog box.

Figure 14-24 Picking Planes for Defining a Parallel Mate (*Parallel3*)

The next step is to create a connection between the tire and the profile cam. However, in *SOLIDWORKS*, it is not allowed to define a cam mate in which the axis of the cylindrical surface of the follower is not parallel with the axis of the cam. Therefore, we need to bring in an auxiliary component, named *wheel_ref*, which acts as the follower. The cam mate can be defined between *wheel_ref* and the profile cam, meanwhile *wheel_ref* is connected to the wheel at a given point so that the motion due to cam rotation can be transmitted to the wheel through *wheel_ref*.

Before defining a cam mate, we suppress two mates to provide adequate freedom and avoid conflict in the assembly. These two mates are *Angle1 (fr_rocker_r1<1>,shock_lower<1>)* and *Parallel3 (HARDPOINTS<1>,profile<1>)*. In the *FeatureManager* tree (or *SOLIDWORKS* browser), expand the *Mates* node 🔗. Right click *Angle1* and choose *Suppress*, and then right click *Parallel3* and choose *Suppress*.

From the pull-down menu, choose *Insert > Mate*. In the *Mate* dialog box, click the *Mechanical Mates* tab and click *Cam* ⟐. In the graphics area, select a surface in *profile* for the *Cam Path* box, and then move the pointer closer to the outer surface of the tire until a label *Boss-Extrude of WHEEL_REF<1>* appears (see Figure 14-25). Note that you may need to right click *WHEEL_REF* and select *Show* to display it. Click the surface for *Cam Follower* box; *Face<3>@WHEEL_REF-1* appears in the *Cam Follower* box. In the graphics area, the profile and *wheel_ref* should come into contact like that of Figure 14-26. Click the checkmark ✔ twice to close the dialog box (click *OK* if a warning message appears).

Creating a Motion Model

Now the model has been completely assembled. The next step is to set up a kinematic analysis using *SOLIDWORKS Motion*. Click the *Motion Study 1* tab at the bottom of the graphics area to start a new study. Right click the tab and rename the motion study as *Kinematic*.

To add a rotary motor that drives the cam, click the *Motor* 🔄 button from the *Motion* toolbar to bring up the *Motor* dialog box. Select *Rotary Motor* 🔄, then pick the *A_1* axis of the profile cam (*profile<1>*) for *Component/Direction*.

Figure 14-25 Creating a Cam-Follower Assembly
Mate between *profile* and *wheel_ref*

Figure 14-26 *profile* and *wheel_ref*
properly assembled

Figure 14-27 Defining a Rotary Motor

A red circular arrow showing the rotational direction will appear, as shown in Figure 14-27. Click ⟳ to switch the direction if necessary. Choose *Constant speed* and enter *60* RPM as the rotational speed for the cam. Click the ✔ button on top of the dialog box to accept the motor definition. *RotaryMotor1* can now be found in the *MotionManager* tree.

Defining and Running Simulation

Choose *Motion Analysis* for the study. Click ⚙ from the *Motion* toolbar to edit the properties of the study. Enter *500* for *Frames per second* then click ✔ button to accept. Define the duration by dragging the end time key to the *2*-second mark in the timeline area, so that the profile cam will make two complete turns during the simulation. Click the *Calculate* 🖩 button from the *Motion* toolbar to start the simulation. Note that you may need to re-suppress the mates, *Angle1* and *Parallel3*, to allow adequate freedom for the system to move. You should see the cam rotating and the wheel being pushed and pulled by the cam. The cam will stop after two complete turns. Click the *Play from Start* ▮▶ button to replay the animation. Also, you may click 🎞 from the *Motion* toolbar to save the animation as an AVI video.

Displaying Simulation Results

Next, we will create graphs of the simulation results to monitor the three measures: vertical wheel travel, shock travel, and camber angle.

Click the *Results and Plots* 📉 button from the *Motion* toolbar. In the *Results* dialog box, choose *Displacement/Velocity/Acceleration*, select *Linear Displacement*, and then *Z Component*. In the browser, pick reference point *Measure* in *wheel<1>* (expanding *wheel_r1.SLDPRT* in the browser) and the origin of *HARDPOINTS<1>*. Click the checkmark ✔ to accept the definition, and you should see a graph similar to Figure 14-28. If the scales of the *Y* axis are integers, double click on the *Y* axis scales and the *Format Axis* dialog box should appear. Switch to the *Number* tab, change the *Decimal Places* to *1* and click *OK*.

Measure@ wheel_r1

Origin@HARDPOINTS

Figure 14-28 Vertical Position of Wheel Center

This graph shows the vertical position of the wheel center as the profile cam rotates. The center of the wheel travels vertically between about 9.5 in. and 12.6 in. The flat portion (11.2 in.) resembles the flat road profile. The distance between the peak and the flat portion is about 1.4 in. due to the 1.35 in. hump. Similarly, the distance between the flat portion and the crest is about 1.7 in. due to the 1.65 in. ditch.

Close *Plot1*. In the *MotionManager* tree, you can find the graph by expanding the *Results* folder. Click the name of the graph and press the *F2* key to rename the graph as *Vertical_Wheel_Travel*.

The graph of the other two measures, shock travel and camber angle, can be defined following similar steps. To plot the shock travel, Click ⬚, select *Displacement/Velocity/Acceleration* in the *Results* dialog box, choose *Linear Displacement*, and then *Magnitude*. Pick reference point *SHOCKER_FRAME* in *HARDPOINTS<1>* and reference point *ROCKER_SHOCKER* in *rocker_redesign<1>* (within the assembly *fr_rocker<1>*) from the browser to define the shock travel. Click ✔ to create the graph, like that of Figure 14-29, which shows that the shock travels between about 5.9 in. and about 8.4 in. The overall travel distance is about 2.5 in., which is probably too large for such a small hump or ditch. In fact, in the simulation, it appears that the shock is compressed too much, in which the piston penetrates into its reserve tube. In reality, this will not happen. However, the simulation raises a flag indicating that there could be severe contact within the shock, leading to potential part failure. Rename the graph as *Shock_Travel* in the *MotionManager* tree.

Figure 14-29 Distance of Shock Travel

The last measure is the camber angle of the wheel. A camber angle is the angle made by the wheel of a ground vehicle; specifically, it is the angle between the vertical axis of the wheel and the vertical axis of the vehicle when viewed from the front or rear. It is used in the design of steering and cornering. If the top of the wheel is further out than the bottom (that is, away from the axle), it is called a positive camber; if the bottom of the wheel is further out than the top, it is called negative camber. In this model, the camber angle will be defined as the rotation angle of the upright along the *X*-axis of the reference triad.

Click the *Results and Plots* ⬚ button to bring up the *Results* dialog box. Choose *Other Quantities*, *Euler Angles*, and *Phi*. For more information regarding the Euler angles, please refer to resources such as Dynamics textbooks or Wikipedia (en.wikipedia.org/wiki/Euler_angles). In the graphics area, pick the flat surface on *RH_upright<1>* as shown in Figure 14-30 and click the ✔ button. As shown in Figure 14-30, the camber angle was set to be about *–1* degree on the flat terrain. The camber angle varies to *–2.5* and *0.4* degrees, respectively, when the tire goes over the hump and the ditch. In general, camber angle alters the handling qualities of a particular suspension design. In general, negative camber improves grip when cornering. This is because it places the tire at a more optimal angle to the road, transmitting the forces through the vertical plane of the tire, rather than through a shear force across it. However, excessive negative camber change in the hump can cause early lockup under breaking or wheel spin under acceleration. There is only limited information that can be obtained by conducting kinematic analysis of the quarter suspension. Ultimately, a full-vehicle dynamic simulation must be carried out to fully understand the suspension design and hopefully develop a strategy for design improvement.

Interference Detection

Before completing the lesson, we want to check if there is any interference between parts when the suspension is in motion. However, there are too many parts in this assembly. The interference checking capability we learned in *Lesson 7* requires more time for *Motion* to perform the calculation. Instead, we use the interference detection capability provided in the standard assembly mode of *SOLIDWORKS* to just identify if there is any major concern in part penetration at the current configuration.

From the pull-down menu, choose *Tools > Evaluate > Interference Detection*. In the *Interference Detection* dialog box (Figure 14-31), click the *Calculate* button. Interferences detected are listed in the *Results* area. Expand interferences detected to learn more, for example, *Interference3*, and click parts with interference in between to see them in the graphics area (Figure 14-31). Note that some of the interferences detected may be simply modeling errors. However, some may reveal significant design defects that need to be taken care of before entering the next design phase or manufacturing.

Figure 14-30 Camber Angle

Figure 14-31 Interference Detection

Full-Vehicle Dynamic Simulation

A final note: a full-vehicle dynamic simulation model was created in *ADAMS/Car*, using the model templates provided, as shown in Figure 14-32a. With *ADAMS/Car*, users can simply enter vehicle model data into the templates, and *ADAMS/Car* will automatically construct subsystem models, such as engine, shock absorbers, tires, as well as the full vehicle assemblies. Once these templates are created, they can be made available to novice users, enabling them to perform standardized vehicle maneuvers. The vehicle model was then simulated for various test scenarios, including skid pad racing, which is a constant radius cornering simulation as shown in Figure 14-32b.

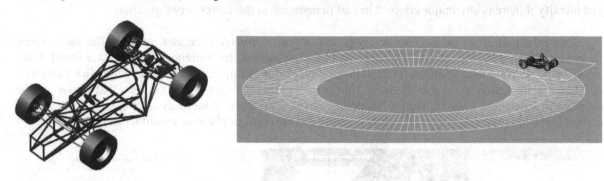

(a) 15 DOFs *ADAMS/Car* Model   (b) Skid Pad Racing

Figure 14-32 Vehicle Dynamic Simulation of Formula SAE Racecar

APPENDIX A: DEFINING JOINTS

Degrees of Freedom

Understanding degrees of freedom is critical in creating successful motion models. The free degrees of freedom of a mechanism represent the number of independent parameters required to specify the position, velocity, and acceleration of individual rigid bodies in the system for any given time. A completely unconstrained body in space has six degrees of freedom: three translational and three rotational. If you add a joint, e.g., a revolute joint to the body, you restrict its movement to rotation about an axis, and the free degrees of freedom of the body are reduced from six to one.

For a given motion model, you can determine its number of degrees of freedom using the Gruebler's count. The mechanism's Gruebler count is calculated using the mechanism's total number of bodies, and the numbers and types of joints. As mentioned above, each movable body introduces six degrees of freedom. Joints added to the mechanism constrain the system or remove DOFs. Motion drivers or forces applied to the system remove additional DOFs.

SOLIDWORKS Motion uses the following equation to calculate the Gruebler's count:

$$D = 6M - N - O \qquad\qquad (A.1)$$

where D is the Gruebler count representing the total free degrees of freedom of the mechanism, M is the number of bodies excluding the ground body, N is the number of DOFs restricted by all joints, and O is the number of the motion inputs (such as a linear or rotary motor) added to the system.

For kinematic analysis, the Gruebler's count must be equal to or less than *0*. The *ADAMS/Solver* recognizes and deactivates redundant constraints during motion simulation. For a kinematic analysis, if you create a model with a Gruebler's count greater than *0* and try to simulate it, the simulation will not run and an error message will appear.

If the Gruebler's count is less than zero, the solver will automatically remove redundancies. For example, you may apply this formula to a door model that is supported by two hinges modeled as revolute joints. Since a revolute joint removes five DOFs, the Gruebler's count becomes:

$$D = (6 \times 1) - (2 \times 5) = -4.$$

The calculated degrees of freedom result is *–4*, which include five redundant DOFs.

Redundancy

Redundancies are duplicate DOFs constrained by joints. When a joint constrains the mechanism in exactly the same way as another joint (like the door hinge example), the mechanism contains excessive DOFs, also known as redundancies. A joint becomes excessive when it does not introduce any further restriction on a body's motion.

It is important that you understand the redundancies in the motion model and be mindful in reviewing dynamic simulation results at these redundant DOFs. For example, if you model a door using two revolute joints for the hinges, the second revolute joint does not contribute to constraining the door's motion. *Motion* solver detects the redundancies and ignores one of the resolute joints in its analysis. The outcome on the redundant DOFs may be incorrectly reported in reaction results, yet the motion is correct.

For a kinematic simulation where you are interested in displacement, velocity, and acceleration, redundancies in your model do not alter the performance of the mechanism.

You can eliminate or reduce the redundancies in your model by carefully choosing joints. These joints must be able to restrict the same DOFs but not duplicate any, introducing redundancies. After you decide which joints you want to use, you can use the Gruebler's count to calculate the DOFs and check redundancies.

Also, after carrying out a motion analysis, *Motion* will show the number of redundancies in the *Mates* node of the *MotionManager* tree, as shown in Figure A-1. You may ask *Motion* to show you the motion analysis message by selecting *Show all Motion Analysis messages* in the *Motion Study Properties* dialog box, as shown in Figure A-2. Click the *Calculate* button in the *Motion* tool menu to bring up the message window. In the message displayed, *Motion* shows you which redundant DOFs are removed in analysis (please see Figure A-3 as an example).

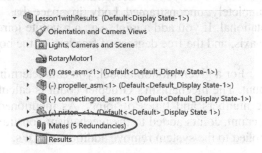

Figure A-1 The *MotionManager* Tree

Figure A-2 The *Motion Study Properties* Dialog

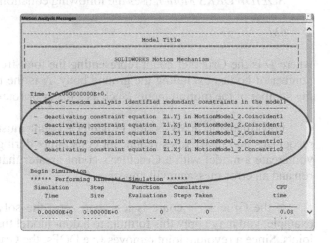

Figure A-3 The Motion Analysis Messages

Joint Types

Before you select a joint to add to your model, you should know what movement you want to restrain for the body and what movement you want to allow. Table A-1 describes the commonly employed joint types in motion simulations and the degrees of freedom they remove.

The following provides more details about the joints listed in the table.

Revolute Joint

A revolute joint, as depicted in Figure A-4, allows the rotation of one rigid body with respect to another rigid body about a common axis. The origin of the revolute joint can be located anywhere along the axis about which the bodies can rotate with respect to each other. The orientation of the revolute joint defines the direction of the axis about which the bodies can rotate with respect to each other. The rotational axis of the revolute joint is parallel to the orientation vector and passes through the origin.

Table A-1 Common Joints Employed for Motion Simulations

Joint Type	DOF Removed			Remarks
	Translation	Rotation	Total	
Revolute	3	2	5	Rotates about an axis
Translational	2	3	5	Translates along an axis
Cylindrical	2	2	4	Translates along and rotates about an axis
Spherical	3	0	3	Rotates in any direction
Universal	3	1	4	Rotates about two axes
Screw	0.5	0.5	1	Coupled rotation and translation along one axis
Planar	2	1	3	Bodies connected by a planar joint move in a plane with respect to each other. Rotation is about an axis perpendicular to the plane.
Fixed	3	3	6	Glues two parts together. Parts constrained by a rigid connection constitute a single body.

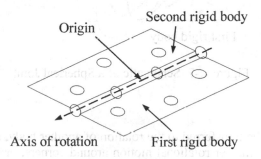

Figure A-4 Schematic of a Revolute Joint

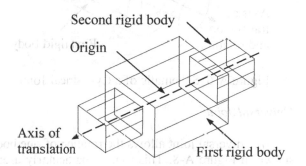

Figure A-5 Schematic of a Translational Joint

Translational Joint

A translational joint allows one rigid body to translate along a vector with respect to a second rigid body, as illustrated in Figure A-5. The rigid bodies may only translate, not rotate, with respect to each other.

The location of the origin of a translational joint with respect to its rigid bodies does not affect the motion of the joint but does affect the reaction loads on the joint.

The orientation of the translational joint determines the direction of the axis along which the bodies can slide with respect to each other (axis of translation). The direction of the motion of the translational joint is parallel to the orientation vector and passes through the origin.

Cylindrical Joint

A cylindrical joint allows both relative rotation and relative translation of one body with respect to another body, as shown in Figure A-6. The origin of the cylindrical joint can be located anywhere along the axis about which the bodies rotate or slide with respect to each other.

Orientation of the cylindrical joint defines the direction of the axis about which the bodies rotate or slide along with respect to each other. The rotational/translational axis of the cylindrical joint is parallel to the orientation vector and passes through the origin.

Spherical Joint

A spherical joint allows free rotation about a common point of one body with respect to another body, as depicted in Figure A-7. The origin location of the spherical joint determines the point about which the bodies pivot freely with respect to each other.

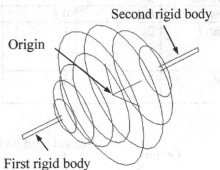

Figure A-6 Schematic of a Cylindrical Joint   Figure A-7 Schematic of a Spherical Joint

Universal Joint

A universal joint allows the rotation of one body to be transferred to the rotation of another body, as shown in Figure A-8. This joint is particularly useful to transfer rotational motion around corners, or to transfer rotational motion between two connected shafts that are permitted to bend at the connection point (such as the drive shaft on an automobile).

The origin location of the universal joint represents the connection point of the two bodies. The two shaft axes identify the center lines of the two bodies connected by the universal joint.

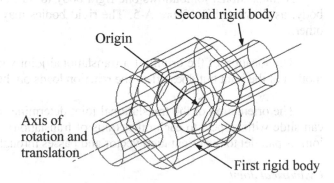

Figure A-8 Schematic of a Universal Joint   Figure A-9 Schematic of a Screw Joint

Screw Joint

A screw joint removes one degree of freedom. It constrains one body to rotate as it translates with respect to another body, as shown in Figure A-9.

When defining a screw joint, you can define the pitch. The pitch is the amount of translational displacement of the two bodies for each full rotation of the first body. The displacement of the first body relative to the second body is a function of the body's rotation about the axis of rotation. For every full rotation, the displacement of the first body along the translation axis with respect to the second body is equal to the value of the pitch.

Very often, the screw joint is used with a cylindrical joint. The cylindrical joint removes two translational and two rotational degrees of freedom. The screw joint removes one more degree of freedom by constraining the translational motion to be proportional to the rotational motion.

Planar Joint

A planar joint allows a plane on one body to slide and rotate in the plane of another body, as shown in Figure A-10.

The orientation vector of the planar joint is perpendicular to the joint's plane of motion. The rotational axis of the planar joint, which is normal to the joint's plane of motion, is parallel to the orientation vector.

Figure A-10 Schematic of a Planar Joint   Figure A-11 Schematic of a Fixed Joint

Fixed Joint

A fixed joint locks two bodies together so they cannot move with respect to each other. The schematic of a fixed joint is shown in Figure A-11.

Notes:

APPENDIX B: THE UNITS SYSTEM

The mass unit lb_m is not quite common to many engineers. The basic physical quantities involved in determining a units system are length, time, mass, and force. These four basic quantities are related through Newton's second law,

$$F = ma \tag{B.1}$$

where F, m, and a are force, mass, and acceleration (length per second square), respectively.

When the unit lb_m for mass is employed, the force unit will be determined by length (in.), mass (lb_m), and second (sec) through Eq. B.1, i.e.,

$$1\ lb_m\ in/sec^2\ (force) = 1\ lb_m\ (mass) \times 1\ in/sec^2\ (acceleration) \tag{B.2}$$

where the force unit, $lb_m\ in/sec^2$, is a derived unit.

From Eq. B.2, a 1 lb_m in/sec^2 force will generate a 1 in/sec^2 acceleration when applied to a 1 lb_m mass block, as shown in Figure B-1a. The same block will weigh 1 lb_f on earth (see Figure B-1b), where the gravitational acceleration is assumed 386 in/sec^2, i.e.

$$1\ lb_f\ (force) = 1\ lb_m\ (mass) \times 386\ in/sec^2\ (acceleration) \tag{B.3}$$

| $F = 1\ lb_m\ in/s^2 = 1\ lb_m \times 1\ in/s^2$ | $F = 1\ lb_f = 1\ lb_m \times 386\ in/s^2$ | $F = 1\ lb_f = 1\ lb_f\ s^2/in \times 1\ in/s^2$ |
| $W = 1\ lb_f = 1\ lb_m \times 386\ in/s^2$ | $W = 1\ lb_f = 1\ lb_m \times 386\ in/s^2$ | $W = 1\ lb_f\ s^2/in \times 386\ in/s^2 = 386\ lb_f$ |

(a) A 1 lb_m in/sec^2 Force Applied to a 1 lb_m Mass Block   (b) A 1 lb_f Force Applied to a 1 lb_m Mass Block   (c) A 1 lb_f Force Applied to a 1 lb_f sec^2/in Mass Block

Figure B-1 Forces Applied on Blocks of Different Masses

Therefore, from Eqs. B.2 and B.3, we have 1 $lb_f = 386$ lb_m in/sec^2, and the force in lb_m in/sec^2 unit is 386 times smaller than that of lb_f that we are more used to. When you apply a 1 lb_f force to the same mass block, it will accelerate 386 in/sec^2, as shown in Figure B-1b.

On the other hand, we have the mass unit, 1 $lb_m = 1/386$ lb_f sec^2/in. It means that a 1 lb_m mass block is 386 times smaller than that of a 1 lb_f sec^2/in block. Therefore, a 1 lb_f sec^2/in block will weigh 386 lb_f on earth. When applying a 1 lb_f force to the mass block, it will accelerate at a 1 in/sec^2 rate, as illustrated in Figure B-1c.

The mass unit slug that we are more familiar with is defined as

$$1\ lb_f\ (force) = 1\ slug\ (mass)\ \times 1\ ft/sec^2\ (acceleration) = 386\ lb_m\ in/sec^2 = 32.2\ lb_m\ ft/sec^2 \qquad (B.4)$$

Therefore, we have 1 slug $= 32.2$ lb_m.

APPENDIX C: IMPORTING *Creo* PARTS AND ASSEMBLIES

From time to time when you use *Motion* for simulations, you may encounter the need for importing solid models from other CAD software, such as *Creo*. *SOLIDWORKS* provides an excellent capability that supports importing solid models from a broad range of software and formats, including Parasolid, ACIS, IGES (Initial Graphics Exchange Standards), STEP (STandard for Exchange of Product data), *SolidEdge*, *Creo*, etc. For a complete list of supported software and formats in *SOLIDWORKS*, please refer to Figure C-1. You may access this list by choosing *File > Open* from the pull-down menu, and pull down the *Files of type* in the *File Open* dialog box.

In this appendix, we will focus on importing *Creo* parts and assemblies. Hopefully, methods and principles you learn from this appendix will be applicable to importing solid models from other software and formats.

SOLIDWORKS provides capabilities for importing both part and assembly. Users can choose two options in importing a solid model. They are Option 1: importing solid features and Option 2: importing just geometry. Importing solid features may bring you a parametric solid model that you will be able to modify just like a *SOLIDWORKS* part. On the other hand, if you choose to import geometry only, you will end up with an imported feature that you cannot change since all solid features are lumped into a single imported geometry without any solid features or dimensions.

Figure C-1 *Open* Dialog Box

Importing geometry is relatively straightforward. In general, *SOLIDWORKS* does a good job bringing in a *Creo* part as a single imported geometry. In fact, several other translators, such as IGES and STEP, support such geometric translations well. IGES and STEP are especially useful when there is no direct translation from one CAD to another.

Importing solid models with solid features is a lot more challenging because solid features embedded in the part geometry, such as holes, chamfers, etc., must be identified first. In addition, sketches that were employed for generating the solid features must be recovered and the feature types, for instance revolve, extrude, sweep, etc., must be identified. With a virtually infinite number of possibilities in creating solid features, it is almost certain that you will encounter problems while importing solid models with feature conversion. Therefore, if you do not anticipate making design changes in *SOLIDWORKS*, it is highly recommended that you import parts as a single geometric feature.

We will discuss the approaches of importing parts and then importing assemblies. In each case, we will try both options, i.e., importing solid features vs. importing geometry. We will use the gear train example employed in *Lesson 8* as the test case and as an example for illustrations.

The Gear Train Example in *Creo*

The gear train assembly consists of one part and three subassemblies. If you have access to *Creo*, you may want to open the final assembly, *gear_train_final.asm*, to check the assembled gear train shown in Figure C-2. There are four components in this assembly: *gbox_housing.prt*, *gbox_input.asm*, *gbox_middle.asm*, and *gbox_output.asm*. The input and output gear assemblies consist of one gear each, *Pinion 1* and *Gear 2*, respectively. The middle gear assembly has two gears, *Gear 1* and *Pinion 2*. The four spur gears form two gear pairs: *Pinion 1* and *Gear 1*, and *Pinion 2* and *Gear 2*, as illustrated in Figure C-2. *Gear 1* and *Pinion 2* are mounted on the same shaft.

There are 22 distinct parts in this assembly, as listed in Table C-1. All the parts and assemblies can be found in folder Appendix C.

Figure C-2 The Gear Train Assembly

Table C-1 List of Parts and Assemblies in Appendix C Folder

Part/Subassemblies	Part Names	Remarks
gbox_housing.prt		
gbox_input.asm	wheel_gbox_shaft_input.prt	
	wheel_gbox_pinion_1s.prt	Pinion 1
	spacer_12×18×5mm.prt	
	spacer_12×20×1mm.prt	
	bearing_12×18×8mm.prt (2)	
	spacer_10×18×014mm.prt	
	wheel_gbox_sft_mid_washer.prt	
	screw_tapper_head_5×15.prt	
	screw_set_tip_6×6.prt (2)	
gbox_middle.asm	wheel_gbox_pinion_2s.prt	Pinion 2
	wheel_gbox_gear_1s.prt	Gear 1
	wheel_gbox_shaft_mid_pinion.prt	
	wheel_gbox_shaft_mid_gear.prt	
	bearing_12×18×8mm.prt (2)	
	screw_tapper_head_5×28.prt (6)	
	wheel_gbox_sft_mid_washer.prt (2)	
	screw_tapper_head_5×15.prt (2)	
	align_pin_4×27mm.prt (2)	
gbox_output.asm	wheel_gbox_gear_2s.prt	Gear 2
	wheel_gbox_connect_wheel.prt	
	bear_tap_roller25×47×15mm.prt	
	screw_straight_head_4×15.prt (10)	
	align_pin_4×20mm.prt (2)	
	wheel_gbox_connect_wh_setscrew.prt (4)	

Importing *Creo* Parts

We will import the gear housing (*gbox_housing.prt*) shown in Figure C-3 using both options. We will try the first option, i.e., importing solid features. If you have access to *Creo*, you may choose from the pull-down menu *Tools > Model Player* to see the sequence of feature creation. As shown in Figure C-3 (*Creo Model Tree* window), there are 8 datum features and 14 solid features. *SOLIDWORKS* will try to import these 14 solid features from *Creo*.

Figure C-3 Gear Housing Part
in *Creo*

Figure C-5

Figure C-4

Option 1: Importing Solid Features

Start *SOLIDWORKS* and choose *File > Open*. Change to folder *Appendix C* (or the folder where you have these gear train parts and assemblies). In the *File Open* dialog box (Figure C-1), pull down the *Files of type* list, and choose *ProE Part (*prt, , *.prt.*, *.xpr)*. You should see a list of *Creo* parts in the dialog box. Click *gbox_housing.prt*, and click *Open*. In the *Pro/E & Creo to SOLIDWORKS Converter* dialog box (Figure C-4), choose *Analyze the model completely* (default), and click *Import material properties*, *Import sketch/curve entities*, and *Import geometry from hidden sections*. Click *OK*.

Translation Report

Features 14, Recognized 12, Imported 12

Feature	Recognition st...	Import status
1. Extrude1	Yes	Yes
2. Extrude2	Yes	Yes
3. Extrude3	Yes	Yes
4. Extrude4	Yes	Yes
5. Cut-Revolve1	Yes	Yes
6. Cut-Revolve2	Yes	Yes
7. Revolve1	Yes	Yes
8. Sketch8	NO	-
9. Extrude5	Yes	Yes
10. Extrude6	Yes	Yes
11. _Chamfer	NO	-
12. Extrude7	Yes	Yes
13. Fillet1	Yes	Yes
14. Extrude8	Yes	Yes
	12/14	12*/14 (*0 with errors)

Figure C-6

In the next dialog box (Figure C-5), click the *Features* button. The conversion process will start. You will see sketches and solid features appear in the graphics screen. After about a minute, the *Translation Report* dialog box, as shown in Figure C-6, appears, summarizing the results of the translation. The report indicates that 12 out of 14 features were recognized and translated. The converted model and features listed in the browser are shown in Figure C-7.

As shown in Figure C-7, there is one dangling sketch, *Sketch8*, representing the unrecognizable solid feature in addition to the chamfer feature. You may identify the sketch in the graphics area by clicking its name listed in the browser. In addition, the back plate (*Extrude1* in the browser) is recognized incorrectly. Certainly, *SOLIDWORKS* is capable of importing some parts correctly and completely, especially when the solid features are relatively simple (but not this gear housing part).

If you take a closer look at any of the successful solid features, for example *Extrude3*, you will see that the sketches (for example *Sketch3* of *Extrude3*) of the solid features do not have complete dimensions. Usually a *(−)* symbol is placed in front of the sketch, indicating that the sketch is not fully defined.

Apparently, this translation is not satisfactory. Unfortunately, this translation represents a typical scenario you will encounter for the majority of the parts. In many cases, it may take only a small effort to repair or re-create wrong or unrecognized solid features. However, when you translate an assembly with many parts, the effort could be substantial.

Figure C-7

Option 2: Importing Geometry

Importing geometry is more straightforward and has a higher success rate than that of importing solid features.

Repeat the same steps to open the gear housing part *gbox_housing.prt*. In the *Creo to SOLIDWORKS Converter* dialog box (Figure C-8), choose *Import geometry directly* (default), and then *Knitting* (default) in order to import solid models instead of just surface models. Note that if you choose *BREP* (Boundary Representation), only boundary surfaces will be imported. Click *OK*.

Figure C-8

Imported geometry

Figure C-9

The conversion process will start. After about a minute or two, the converted model will appear in the graphics screen, as shown in Figure C-9. In addition, an entity *Imported1* will appear in the browser (Figure C-9). As mentioned earlier, there will be no parametric solid feature with dimensions and sketch converted if you choose *Option 2*. However, the geometry converted seems to be accurate. All the geometric features in *Creo* shown in Figure C-3 were included in this imported feature. This translation is successful. Since we do not anticipate making any change to the gear housing, this imported part is satisfactory. The gear housing part, *gbox_housing.sldprt*, employed in *Lesson 8* was created by using *Option 2*.

Importing *Creo* Assembly

We will import the input gear assembly (*gbox_input.asm*) shown in Figure C-10 using both options. We will try *Option 1* first, i.e., importing solid features. As shown in Figure C-10 (*Creo Model Tree* window), there are 11 parts (and several datum features) in this assembly. *SOLIDWORKS* will try to import this assembly as well as the 11 parts from *Creo*.

Figure C-10 Input Gear Assembly in *Creo*

Figure C-11

Option 1: Importing Solid Features

Repeat the same steps to open the input gear assembly *gbox_input.asm*. In the *Creo to SOLIDWORKS Converter* dialog box (Figure C-11), choose *Use feature import for all parts*, and choose *Overwrite* for *If same name SOLIDWORKS file is found* (just in case you have *SOLIDWORKS* files with the same file names in the same folder). Choose *Import material properties* and *Import sketch/curve entities*. Click *Import*. The conversion process will start.

You will see sketches, solid features, and solid models appear in the graphics screen. After about a minute, the translation process is completed. The converted assembly and the browser with parts listed are shown in Figure C-12.

As shown in Figure C-12, parts are not completely converted. Major solid features are missing, for example, the pinion 1 (*wheel_gbox_pinion_1s<1>*), where most solid features are not converted. If you expand the part, you will see that only two extrude features were converted; the remaining entities are mostly sketches. The remaining parts were converted. However, the *Mates* branch in the browser is completely empty, implying that no assembly mates have been imported.

Apparently, this translation is not satisfactory. A non-trivial effort will have to be spent in order to reconstruct the solid features (therefore, solid models) as well as the final assembly.

Figure C-12

Option 2: Importing Geometry

Importing geometry is also more straightforward for assembly and has a higher rate of success.

Repeat the same steps to open the input gear assembly *gbox_input.asm*. In the *Creo to SOLIDWORKS Converter* dialog box (Figure C-13), choose *Use body import for all parts* (default), and then *Knitting* (default) in order to import solid models. Choose *Overwrite* for *If same name SOLIDWORKS file is found*, and choose *Import material properties* and *Import sketch/curve entities*. Click *Import*. The conversion process will begin.

Figure C-13

Imported geometry

Figure C-14

After about a minute or two, the converted assembly will appear in the graphics screen, as shown in Figure C-14. The assembly and all 11 parts seem to be correctly imported. If you expand any of the part branch, for example, the gear (*wheel_gbox_pinion_1s<1>*), you will see an imported feature listed, as depicted in Figure C-14. Again, there is no solid feature converted in any of the parts. In addition, the *Mates* branch is empty.

Since we do not anticipate making any change to this input gear assembly, this imported assembly is satisfactory, except it does not have any assembly mates. Assembly of all 11 parts (it may be more for some other cases) will take a non-trivial effort. Since we do not anticipate making changes in how these parts are assembled, we will merge all 11 parts into a single part.

In *SOLIDWORKS*, you can join two or more parts to create a new part in an assembly. The merge operation removes surfaces that intrude into each other's space and merges the parts into a single solid volume. We will insert a new part into the assembly and merge all 11 parts into that new part.

Choose from the pull-down menu *Insert > Component > New Part*. A new part with an assigned name is listed in the browser (Figure C-15). *SOLIDWORKS* is expecting you to select a plane or a flat face to place a sketch for the new part.

Figure C-15   New part

Figure C-16

Front Plane

Click a plane or planar face on a component; for example, pick the assembly *Front* plane from the browser. The *Front* plane will appear in the graphics screen (Figure C-16). In the new part, a sketch opens on the selected plane. Close the sketch by clicking the *Close* button ✖ and choose *Discard Changes and Exit*. Because we are creating a joined part, we do not need a sketch.

Next, we will select all 11 parts and merge them into the new part.

From the browser, click the first part *wheel_box_shaft_input<1>*, press the *Shift* key, and then click the last part, *screw_set_tip_6×6<2>*. All 11 parts will be selected.

From the pull-down menu, choose *Insert > Features > Join*. The *Join* window will appear (overlapping with the browser) as shown in Figure C-17. In the *Join* window, all 11 parts are listed. All you have to do is to click the checkmark on top to accept the parts. Save the part (choosing from the pull-down menu *File > Save*). Choose *Save All* in the *Save Modified Documents* dialog box (Figure C-18), and then click *OK* in the *Save As* dialog box (Figure C-19). A part file *gbox_input.sldprt* is created in the folder. Close the entire assembly model.

Now open the part *gbox_input*. Make sure you open *gbox_input.sldprt* instead of *gbox_input.sldasm*. The part *gbox_input* will appear in the graphics area. In addition, all entities belonging to this part will be listed in the browser, as shown in Figure C-20. Note that there is an arrow symbol ->? to the right of the entity, *Join1*. This symbol indicates that these entities enclosed in this part refer to other parts or assembly. Note that the *Join1* branch has the same symbol. Expand the *Join1* node; you will see 11 parts listed, all with arrows, pointing to the actual parts currently in the same folder.

Figure C-17

Figure C-18

Figure C-19

Figure C-20

The three gear parts, *gbox_input.sldprt*, *gbox_middle.sldprt*, and *gbox_output.sldprt*, employed for *Lesson 8* were created following the approach discussed. One axis in each part that passes through the center hole of the gear was created simply by intersecting two planes, for example, *Top* and *Right* planes for the axis in *gbox_input.sldprt*, as shown in Figure C-21. These axes are necessary for creating gear pairs, as discussed in *Lesson 8*.

Figure C-21

Notes: